OXFORD MEDICAL PUBLICATIONS

Human Energy Requirements

HUMAN ENERGY REQUIREMENTS

A Manual for Planners and Nutritionists

W. P. T. JAMES

Professor of Nutrition
Rowett Institute, Aberdeen

and

E. C. SCHOFIELD

Honorary Research Fellow
London School of Hygiene and Tropical Medicine

Published by arrangement with the
Food and Agriculture Organization of the United Nations
by
OXFORD UNIVERSITY PRESS
Oxford   New York   Tokyo
1990

Oxford University Press, Walton Street, Oxford OX2 6DP
Oxford New York Toronto
Delhi Bombay Calcutta Madras Karachi
Petaling Jaya Singapore Hong Kong Tokyo
Nairobi Dar es Salaam Cape Town
Melbourne Auckland
and associated companies in
Berlin Ibadan

Oxford is a trade mark of Oxford University Press

Published in the United States
by Oxford University Press, New York

British Library Cataloguing in Publication Data
James, W. P. T. (William Philip Trehearne)
Human energy requirements.
1. Man energy metabolism
I. Title II. Schofield, E. C.
612'.39
ISBN 0-19-261891-1

Library of Congress Cataloging in Publication Data
James, W. P. T. (William Philip Trehearne), 1938–
Human energy requirements: a manual of calculations and practical
applications/W. P. T. James and C. Schofield.
p. cm.—(Oxford medical publications)
Includes bibliographical references.
1. Energy metabolism. 2. Nutrition policy. 3. Food supply.
I. Schofield, Claire. II. Food and Agriculture Organization of the
United Nations. III. Title. IV. Series.
[DNLM: 1. Energy Metabolism. 2. Mathematics. 3. Nutritional
Requirements. QU 125 J29h]
QP176.J36 1989 363.8'2—dc20
ISBN 0-19-261891-1

Typeset by Cotswold Typesetting Limited, Cheltenham
Printed in Great Britain by
Courier International, Tiptree, Essex

Foreword

P. Lunven, Director, Food Policy and Nutrition Division, FAO

This manual represents a milestone in the continuous involvement and experience of FAO of almost forty years in human energy requirements. The report of the first Committee on Calorie Requirements was published by FAO in 1950. Reports of subsequent committees and expert consultations were published in 1957 (FAO), 1973 (FAO/WHO) and most recently in 1985 under the WHO Technical Report Series (FAO/WHO/UNU). Thus far in the reports emphasis had been placed on methodological progress based on the latest scientific information available with a view to deriving estimates of human energy requirements while less attention was paid on how to apply the requirements to practical food and nutrition planning. However, following the 1973 report it was realized that the practical application of requirements in general and of energy in particular was a complex topic which required serious consideration and deserved to be fully treated in a special report.

Following the publication of the 1985 FAO/WHO/UNU report and on the basis of additional information obtained on certain factors used in calculating requirements, FAO began to make arrangements for preparing a manual aimed at advising and guiding planners and nutritionists, among others, on how to apply the methodology presented in the 1985 report. The present Manual is the result of more than two years of intense work and fruitful cooperation between a group of consultants and FAO staff. It addresses issues identified during discussions with experts at a meeting held in FAO in December 1987 as well as with other potential users of requirements and includes a number of methodological refinements developed by using computer modelling techniques. In the course of the work, a micro computer spreadsheet programme based on the computer modelling was developed. The spreadsheet replicates the calculation steps described in the manual, drawing upon data supplied with the programme. The computer software is included as companion to the manual.

The first part of the report, under Chapter 1, is a general overview of how energy requirement levels affect a wide range of economic and developmental issues. This brief and illustrative section describes summarily the method of calculating energy requirements and predicts the effect of different assumptions on the final energy requirement value.

The remaining seven chapters of the manual present a more detailed discussion addressed to nutritionists and others who wish to examine the basis of energy requirements in greater depth. The manual describes not only the factors used in calculating energy requirements and how to apply the methodology but also provides sets of data needed to apply the methodology. At this stage, we are aware that much more work is needed to document people's activities and their corresponding relative energy costs, as the manual has been prepared on the basis of information currently available in 1985.

The manual is based on the work of W. P. T. James and E. C. Schofield in collaboration with the technical staff of the Nutrition Planning, Assessment and Evaluation Service of the Food Policy and Nutrition Division of FAO and with A. Ferro-Luzzi of the Istituto Nazionale della Nutrizione, Rome. The spreadsheet programme was written at the Rowett Research Institute, Aberdeen, Scotland by T. A. Travis in collaboration with E. C. Schofield with assistance from D. A. Grubb. The User Guide for the spreadsheet was written by J. E. Solesbury and R. C. Weisell of FAO. Collation and editing of the manual was provided by J. H. James.

Professor James, Director of the Rowett Research Institute in Aberdeen, Scotland, was Co-chairman of the Energy Group at the 1981 Expert Consultation which produced the 1985 Report and E. C. Schofield collected, collated, and evaluated additional BMR data from the literature which was used in this manual. Their involvement in the development and finalization of this manual has by far exceeded their original terms of reference as consultants to the activity. We acknowledge with gratitude their unfailing commitment and contribution as well as the contributions of the many others who reviewed and provided comments on the various drafts of the manual.

Contents

Glossary

Main Location

Chapter 1,
Section 1.2

Allowance: the allowance for a nutrient, such as a vitamin or mineral, is a value estimated to cover the needs of 97% of the population. This value is calculated by estimating a mean + 2 SD of the observed requirements in a group of individuals. (See also RDA.) In this manual an allowance of energy is not based on the same system. It is the additional amount of energy to be provided in excess of the *average* energy requirement to cover needs for extra growth or desirable activity. Planning allowances are also specified to cover losses in the food chain.

Chapter 3,
Page 43

Basal Metabolic Rate (BMR): 'the minimal rate of energy expenditure compatible with life'. It is measured under standard conditions of immobility in the fasting state, with an environmental temperature of 26–30°C, which ensures no activation of heat generating processes (e.g. shivering).

Chapter 7,
Page 92

Body Mass Index (BMI): is defined by the following expression:

$$\frac{\text{Weight in kg}}{\text{Height}^2 \text{ in m}}$$

The acceptable range of BMI is: 20.5–25 for men,
and 18.7–23.8 for women.

Chapter 5

Desirable ... standards are those that express a level of body size or energy expenditure which is compatible with the long-term health of all groups in the population. The 1985 WHO/FAO/UNU report specifies energy needs as fulfilling the above criteria. In this manual these desirable standards or allowances have been highlighted, in relation to the lower requirement values.

Chapter 3,
Page 50 and
Appendix 5

Integrated Energy Index (IEI): the cost of an activity or occupation which, in contrast to PAR, includes the time spent pausing whilst conducting the activity. It is calculated by the expression:

$$\frac{\text{The energy cost of an activity for the time allocated}}{\text{The basal metabolic rate for the same period}}$$

Chapter 2,
Page 35

Kilojoule (kJ): a joule is the energy expended when one kilogram is moved 1 metre by a force of 1 Newton. This is the standard unit of energy used in human energetics. Because nutritionists are concerned with large amounts of energy, they conventionally use kilojoules ($kJ = 10^3$ J) or megajoules ($MJ = 10^6$ J). One calorie or kcalorie is equivalent to 4.184 kJ.

Chapter 3,
Page 53
Table 3.4

Physical Activity Level (PAL): total energy requirement for a 24 hour period. It is calculated by the expression:

$$\frac{\text{the total energy required over 24 hours}}{\text{the basal metabolic rate over 24 hours}}$$

Chapter 3,
Page 47 and
Appendix 5

Physical Activity Ratio (PAR): the actual energy cost of an activity, minute by minute. It is calculated by the expression:

$$\frac{\text{the energy cost of an activity per minute}}{\text{the energy cost of the basal metabolic rate per minute}}$$

Chapter 1,
Section 1.2

Recommended Daily Allowance (RDA): for nutrients it has become the convention to add an extra safety margin to estimated vitamin, mineral or protein allowances to ensure that the whole population's needs are covered. RDAs may vary from country to country according to government recommendations. They are also known as Recommended Dietary Intakes (RDIs).

Chapter 1,
Section 1.2

Requirements: for each individual the requirement for a particular nutrient or for energy is the amount necessary to ensure normal physiological functions, and to prevent symptoms of deficiency occurring.

1 An overview of energy requirements and allowances

1.1 Introduction

1.1.1 The use of energy allowance figures

The amount of food needed by a population is one of the basic factors affecting economic, agricultural, developmental and political strategies in many countries. It is therefore important to have a clear picture of the population's needs for food and how they are likely to alter in response to demographic change and to a variety of social and economic adjustments. Figure 1.1 shows the areas of expertise which influence or are affected by energy requirement estimates. Each of the links specified by a bold arrow is illustrated in the following pages.

This manual describes how to calculate the energy *requirements* of a household, a group of people, or a population. Extra individual *allowances* can also be specified if one wishes to provide more food to sustain, for example, better

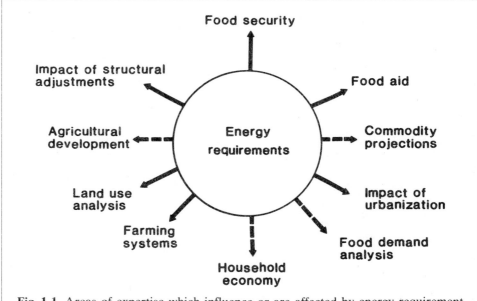

Fig. 1.1 Areas of expertise which influence or are affected by energy requirement estimates.

growth in children or a higher rate of physical activity in adults. These individual allowances can be aggregated, but adjustments are needed at a household, regional, and national level for the various post-harvest losses, e.g. in food production, processing, storage, distribution, and in the preparation of food within the household. This analysis makes the distinction between food requirements and economic indices at a national level easier to understand. It may also focus attention on alternative methods of improving food consumption without the need to increase food production.

1.1.2 Specifying the energy requirements of a group

The energy *requirements* of a group or population are readily predicted. The requirements depend on age, size, sex, and physical activity, as shown in Fig. 1.2.

The energy requirements of a population are fixed and can be specified within ± 5 per cent. The requirements only change when individuals within the population change their activity or size, or when the structure or number of the population alters.

Food demand and food requirements are not the same:

Demand: *a social or economic characteristic*	**Requirement:** *a biological feature*
Variable	Fixed within ± 5 per cent for weight-stable adults with a constant lifestyle
Culturally determined	Biologically determined, e.g. by sex, age, weight, and activity
Economically affected	Occupational change can have an impact by altering physical activity

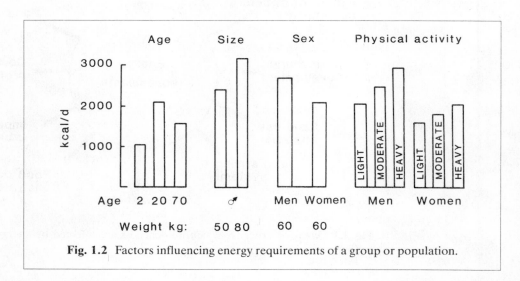

Fig. 1.2 Factors influencing energy requirements of a group or population.

Are food needs defined in socio-economic (demand) terms or as the body's need for food (requirements)?

Children and adults, when provided with unlimited amounts of food, usually eat enough to satisfy their energy requirements. Only when very bulky foods of low energy density are eaten do children find difficulty in eating enough to satisfy their needs. Food requirements are therefore taken normally to reflect energy requirements. The total volume of food consumed varies with the energy density of the diet. The quality of food is also important because a minimum level of nutrients, e.g. vitamins, protein, and minerals, must be obtained from the diet. This manual is concerned with predicting the *total amount of food energy* needed by groups of children and adults; the quality of the food must then be considered separately.

The word 'consumption' can be interpreted in two ways, either in an economic sense, i.e. foods bought, exchanged, etc. or in a nutritional sense, i.e. foods eaten. In this manual the nutritional meaning is used.

Figure 1.3 shows the general food distribution systems; not all factors are included. The different monitoring systems in use are also specified, so that it is clear that they are measuring different parts of the food chain.

1.1.3 Popular misconceptions about energy requirements

Climate

Climate is *not* normally important in determining energy requirements. Any effect, if present, is small and minimized by changes in clothing, housing and the use of external heat. In emergency conditions, however, reduced metabolic rates resulting from starvation and low environmental temperatures may result in the need for an extra allowance of energy above the normal requirement.

Nutritional elasticity is a limited concept

Nutritional elasticity: Change in nutritional demand in a population is dependent on changing stature, weight, population structure, occupational, and other physical activities. Metabolic elasticity is negligible except in an emergency.

Demand elasticity: Change in demand for 'inferior' or 'luxury' goods is dependent on 'effective' income: food necessities demonstrate limited demand elasticity, i.e. demand is relatively insensitive to income change.

Energy requirement figures should be used to assess the validity of projected commodity demands, these being aggregated and expressed in terms of calories per head per day. Projected commodity demands simply reflect predictions based on national statistics and may bear no relationship to potential changes in energy requirements.

An example of the difference between food supply and energy requirements

Changing patterns of supply and energy requirements for Mexico in the period 1965–1980 are shown in Fig. 1.4.

Food supply per caput has increased, but requirement remains stable because there is a rise in average body weight but a fall in activity levels in the population.

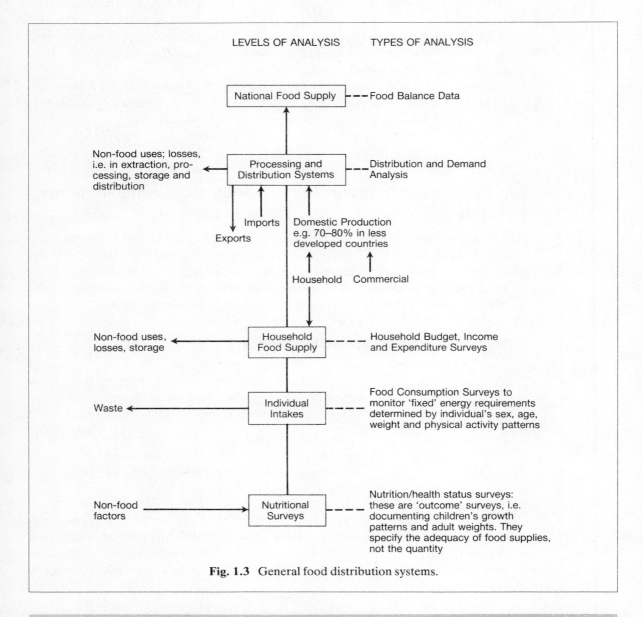

LEVELS OF ANALYSIS   TYPES OF ANALYSIS

National Food Supply – – – Food Balance Data

Non-food uses; losses, i.e. in extraction, processing, storage and distribution ← Processing and Distribution Systems — Distribution and Demand Analysis

Imports
Exports
Domestic Production e.g. 70–80% in less developed countries
Household   Commercial

Non-food uses, losses, storage ← Household Food Supply – – – Household Budget, Income and Expenditure Surveys

Waste ← Individual Intakes – – – Food Consumption Surveys to monitor 'fixed' energy requirements determined by individual's sex, age, weight and physical activity patterns

Non-food factors → Nutritional Surveys – – – Nutrition/health status surveys: these are 'outcome' surveys, i.e. documenting children's growth patterns and adult weights. They specify the adequacy of food supplies, not the quantity

Fig. 1.3 General food distribution systems.

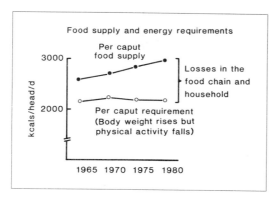

Fig. 1.4 Changing patterns of supply and energy requirements for Mexico, 1965–1980.

The size of the gap may reflect food losses but estimates of per caput food supply are often imprecise. The minimum size of the gap necessary to ensure that vulnerable sections of society have an adequate supply of food is unknown, and will vary from one society to another.

Sources of calories do not affect energy requirements

Sources of calories for Mexico during the period 1965–1980 are shown in Fig. 1.5. Note that data on food supplies are often used to predict the demand for total food and specific commodities. If the population's energy requirements are met then the consumption of a variety of food sources usually ensures that other nutritional needs are met.

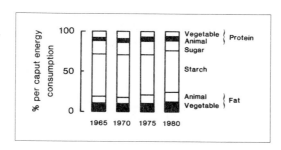

Fig. 1.5 Sources of calories for Mexico, 1965–1980.

1.1.4 Specifying energy allowances for planners

Energy allowances have to be specified either at an individual level or at different levels in the food chain. The allowances at a household or national level take account of food losses and non-food uses. Figure 1.6 is the reverse of Fig. 1.3 and shows how one can build up from individual data a composite set of allowances at a national level.

These are allowances for *planning purposes* and are different from the nutritional allowances specified for protein, minerals and vitamins where amounts are proposed to satisfy everybody's needs, not the average needs as with energy (see 1.2).

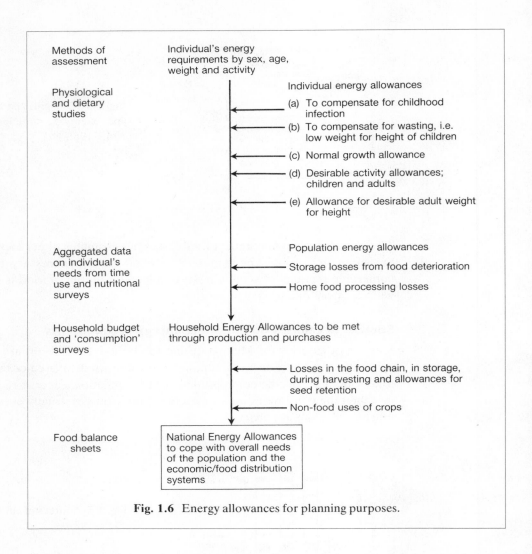

Methods of assessment	Individual's energy requirements by sex, age, weight and activity	
Physiological and dietary studies		Individual energy allowances
	←	(a) To compensate for childhood infection
	←	(b) To compensate for wasting, i.e. low weight for height of children
	←	(c) Normal growth allowance
	←	(d) Desirable activity allowances; children and adults
	←	(e) Allowance for desirable adult weight for height
Aggregated data on individual's needs from time use and nutritional surveys		Population energy allowances
	←	Storage losses from food deterioration
	←	Home food processing losses
Household budget and 'consumption' surveys	Household Energy Allowances to be met through production and purchases	
	←	Losses in the food chain, in storage, during harvesting and allowances for seed retention
	←	Non-food uses of crops
Food balance sheets	National Energy Allowances to cope with overall needs of the population and the economic/food distribution systems	

Fig. 1.6 Energy allowances for planning purposes.

1.1.5 Land use analysis: potential population supporting capacities

Can the available land sustain the projected population? Figure 1.7 shows estimated crop yields for differing input levels of technology. Output depends not only on soil, weather, choice of crops, irrigation, pest control, and management techniques as well as human resources which include skills and adaptability, but also the number of people, their physical capacity and time available to devote to agricultural production.

On the left of Fig. 1.8 is shown a varying production rate which depends both on the length of the growing period and the level of input to the land. For comparison, on the right is displayed the number of calories needed per year to support 1–4 people per 10 hectares in two population types: (a) a population with a predominance of tall, heavy but active adults and few children and (b) a

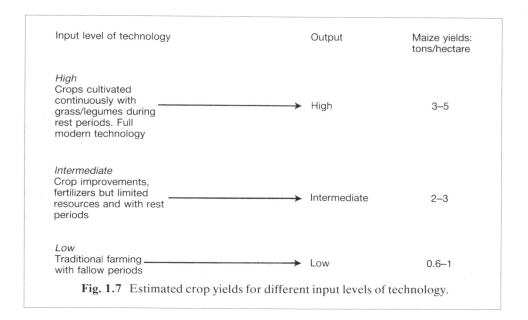

Input level of technology	Output	Maize yields: tons/hectare
High Crops cultivated continuously with grass/legumes during rest periods. Full modern technology	High	3–5
Intermediate Crop improvements, fertilizers but limited resources and with rest periods	Intermediate	2–3
Low Traditional farming with fallow periods	Low	0.6–1

Fig. 1.7 Estimated crop yields for different input levels of technology.

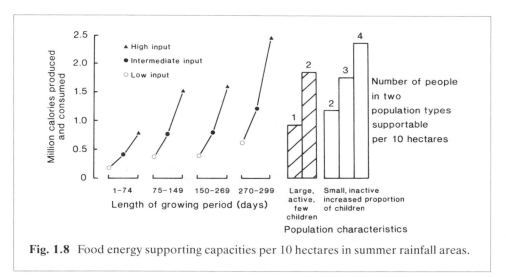

Fig. 1.8 Food energy supporting capacities per 10 hectares in summer rainfall areas.

population with small adults, many children, and an inactive lifestyle. This illustrates the range of nutritional demand in the same terms as agricultural production.

Not included in these analyses is provision for the energy needed in the production of fuel, e.g. wood for cooking, the energy needed for the feeding of draught animals, and the energy used in animal production. Other energy losses, e.g. during the milling and processing of cereals, as well as storage losses at the household level, must be taken into account in the overall planning process.

The ability of land to produce food is limited and the energy requirement of individuals is fixed unless they change their weight or alter their pattern of physical activity. With time, populations show changes in stature, in the growth of children, in average adult body weight, and in physical activity: all these affect energy requirements. The maximum range in energy requirement is displayed both in calories consumed per year and in terms of the supportable populations per 10 hectares of land in this particular climate and soil conditions.

Estimates of the population's energy needs can be obtained in simple form from section 1.3, with options for changing the allowances being explained in section 1.2 and their effects summarized in section 1.5.

1.1.6 National and household food security

One aspect of food security is the ability to cope with shortages by purchasing food or releasing it from food stores and this can be important both at the national and the household level. Food security reserves, specified as a percentage of annual energy needs and often expressed as staple food equivalents, may be established to ensure that food is available.

The size of such food security reserves may be affected more by the cycle of food production and by population numbers than by variations in the population's actual energy requirements (Fig. 1.9).

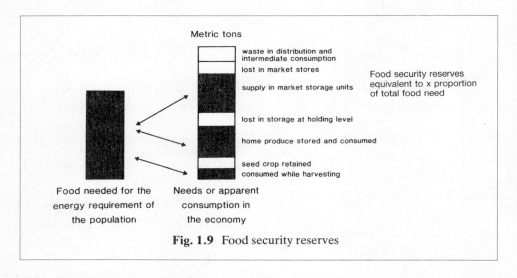

Fig. 1.9 Food security reserves

Countries may improve food security at a national level by having abundant stores; poor, less developed countries may have limited national or regional stores. In any event, the rural population must usually depend upon their own storage systems. These household stores are often large enough to withstand two crop failures in parts of the world where harvests are uncertain. Economic consumption figures before and after a famine are frequently distorted by fluctuations in food stores at the household level. After a famine, food is often stored once supplies become available, so economic consumption figures can rise sharply.

Enhancing national and household food security

The current understanding of food security is based on the interrelated concepts of stable agricultural production and adequate access to food. The broad concept of food security provides a useful framework for attempts to alleviate hunger and malnutrition at either national or household levels.

One measure of the degree of food security that either nations or households enjoy is how successful they are in coping with shortages or interruptions in their food supply. Disrupted supplies may be caused by any factor that interferes with the ability of people to acquire food through normal channels, but in general they are due to: (1) decreases in local food production; (2) losses of purchasing power or exchange entitlements; and/or (3) breakdowns in marketing or distribution systems.

One of the most effective mechanisms for assuring the continued availability of food during such times is the maintenance of reserve stocks of food or other assets. Many governments, in order to enhance food security at the national level, have established Strategic Grain Reserves for emergencies, which are maintained in addition to the normal working stocks, and usually specified as a portion of the national annual cereal needs. When calculating such cereal needs, the nutritional needs of a population, and specifically its energy requirements, will be the starting point. Allowances for losses incurred in storage, processing and distribution are then added.

Villages and households may also maintain reserves, especially if they are primarily dependent upon their own production for their food needs and when national or district/regional storage is limited. While the actual size of a given reserve at any level depends on the unique factors of the relevant food production, supply, storage, demand, and distribution systems, the primary objective is to ensure that adequate stocks are available to meet the nutritional and energy needs of the population for a specified period of time.

The procedures described in this manual are intended to make estimating these needs simple and straightforward.

Approaches to food security by different nations

Figure 1.10 presents an hypothetical approach to food security to illustrate some of the possible effects of different economic systems. On the left the total food stores in a subsistence economy comprise both national and household food stores, with the household stores making a substantial contribution. One would expect that countries with a short growing season, i.e. a low value on the ordinate, would take additional care to ensure adequate stores, since a single crop failure would have serious consequences unless food was available to support the population for many months, if not a year. Evidence from areas of Africa that are vulnerable to crop failure, suggests that households may attempt to store up to 2 years' food supplies. The right hand diagram shows that in affluent societies household food stores are very small. Some countries with the necessary economic and administrative support still ensure very large stocks of

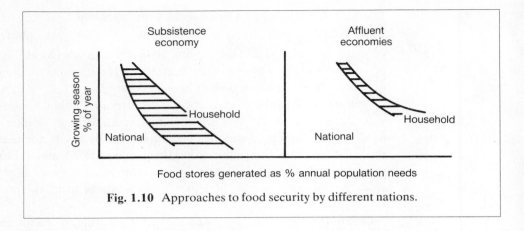

Fig. 1.10 Approaches to food security by different nations.

food as a public policy, even though they have the purchasing power to buy food from other countries should the need arise.

1.1.7 Farming systems and household needs

Farm systems analysis (Fig. 1.11) is a practical technique for improving the analytical accuracy when assessing the performance of farming strategies for rural development. These assessments then affect agricultural development policy. This holistic approach looks at development opportunities at the level of

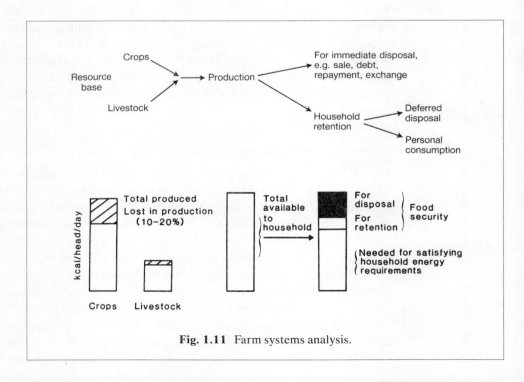

Fig. 1.11 Farm systems analysis.

the small family farm and community. Unpaid family labour is common, off-farm activities are of considerable importance and the preoccupation of the households is to meet subsistence requirements and minimize risks of food shortages. In this context subsistence food requirements are a function of dietary energy needs. Attempts to implement technical, socio-economic, and institutional improvements must be considered within the context of the food needs of the farmer's family and their physical activity patterns. This is a very different problem from those farming systems where labour is employed and where development is often aimed at improving profits.

Out of the total food available, the amount that a farmer needs to retain both for consumption and for postponed sale depends both upon the household's ability to cope with crises, and the family's dietary energy needs. The dietary energy needs of the family reflect the number of household members and their age, sex, and weight, and the physical activity demands of the socio-agricultural system, i.e. who does what and for how long.

1.1.8 The impact of urbanization on food demand

FAO studies have highlighted the importance of urbanization as one of the major factors influencing patterns of food demand.

Time trends in demand for food expressed as the z factor reflect among other factors the influence on demand of 'structural' changes such as changes in income distribution, consumer preferences, increased market penetration of specific food commodities and special entitlements to free or subsidized foods. These factors are in part affected by urbanization itself. FAO studies have highlighted the importance of urbanization as one of the major forces associated with changes in the patterns of food demands.

Household surveys often show differences in food purchases and consumption patterns in urban and rural areas in less developed countries with increases in the consumption of sugar, alcohol, soft drinks, highly milled cereals, and processed foods in the urban communities[1]. Although there is usually an increase in fat and animal protein consumption in urban areas, the total protein intake may decline and energy intakes are almost invariably lower in urban communities, as the data in Table 1.1 suggest.

Table 1.1 Impact of urbanization on food demand

Examples	Fewer calories per caput/day		Variable protein g/day		More animal protein g/day		More fat g/day	
	Rural	Urban	Rural	Urban	Rural	Urban	Rural	Urban
Brazil	2640	2428	79.2	74.0	29.2	30.7	60.0	63.0
Bangladesh	2254	1732	57.4	49.5	7.9	12.1	17.2	25.0
Korea, Repub. of	2181	1946	66.9	62.8	5.1	10.9	15.8	19.5
Trinidad & Tobago	3011	2850	81.7	83.6	31.8	43.3	84.0	95.8

Possible explanations for a fall in household calorie purchases in urban areas include:

1. inadequate income to purchase foods: reduced intake will lead to weight loss and perhaps lower levels of activity;
2. reduction in energy needs of population in response to less physical activity: better transport, occupational change, sedentary recreations;
3. selective transfer to urban area of household members, e.g. children with lower energy requirements;
4. reduced household waste.

The last two explanations appear to be unlikely on the basis of survey data, which show more adult males migrating to urban areas and more rather than less household waste.

With knowledge of the distribution of sex, age, weight, and activity patterns in the urban population obtained from household surveys, a better understanding of the reasons for these observations will be possible. The use of human energy requirement figures derived from such information will permit the food requirements of the cities and urban areas to be determined with precision.

1.1.9 Assessing the effects of structural adjustment

A variety of political decisions and social and economic changes may affect the availability and cost of food (Fig. 1.12). Some countries are having to cope with such effects as they are undertaking adjustments to their economic and social policies to comply with specific fiscal requirements which are a pre-requisite for the release of additional foreign loans (e.g. the World Bank's Structural Adjustment Loans). These changes include currency devaluations, better incentives for export production, fewer government constraints on domestic prices, the

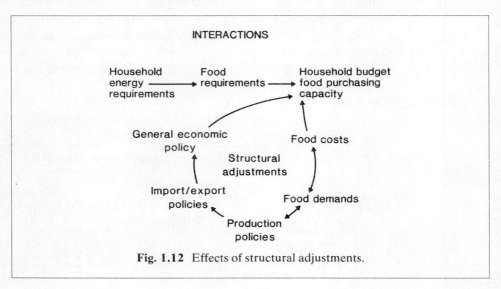

Fig. 1.12 Effects of structural adjustments.

removal of producer and consumer subsidies, and reduced public expenditure on health, education and social services. The net effect of all these changes is to reduce the real incomes of at least some sections of society, if only temporarily.

Social, economic, and climatic factors can influence the food production capacity of vulnerable groups within a population; in urban areas the population is particularly dependent on food purchases and the ability to afford enough food energy. Some groups in a society may therefore become vulnerable to malnutrition because their ability to grow or purchase food is insufficient to meet their energy requirements. Consequently the impact of structural adjustment on the nutrition of the population must be considered.

An analysis of the possible effects of these structural adjustments depends on understanding the impact of changes on the purchasing power and food costs of vulnerable groups. Their energy requirements will remain the same whatever the economic changes unless they become unemployed and sedentary: they then become dependent on food aid or poverty alleviation programmes.

During structural adjustment the population as a whole should be considered as potentially vulnerable, because adult men and women, as well as other groups, will be unable to sustain their production capacity without enough food. Traditionally vulnerable groups are considered by nutritionists to be young children, pregnant and lactating women, and the elderly. Their vulnerability leads to the view that permanent damage may ensue if children or pregnant women are poorly fed. The elderly are thought to have an inadequate ability to obtain an appropriate share of the household's food supply in times of difficulty.

The monitoring of the effects of structural adjustment should include nutritional surveys, e.g. of body weights of adults and children, but estimating the extent of compensatory adjustment in policies requires an understanding of energy requirements.

1.1.10 Food aid

Estimating a country's need for food aid depends on an adequate knowledge of both food needs and food supply. To assess food supply, information is needed on (a) food production, (b) production, distribution and storage losses, (c) fluctuation in food stocks, (d) seed requirements, and (e) food imports (including donations) and exports.

Food aid is required if food needs exceed food supply and additional imports cannot fill the gap. Problems may also exist in the distribution system or in the purchasing power of the population.

Assessing food needs

There are two approaches:

1. *The status quo method.* Annual figures for food availability, even if inaccurate, can indicate trends in food supply, and a falling trend may signal the need for food aid. With the status quo method, average food supply per annum during a base period of 'normal' years is used as a proxy for food needs, 'normal' being

defined as years of so-called 'adequate food production'. This pragmatic approach first equates food availability with food consumption and, secondly, assumes that food availability during the base period is sufficient to meet food needs. In addition to their doubtful quantitative validity, the actual energy values calculated from such statistics have no validity in nutritional or biological terms. Given the uncertainty of the measures of food production, imports, etc., the method does, however, give an indication of possible worsening or improving circumstances.

2. *The nutrition-based method.* The nutrition-based method depends on estimating food needs according to the method in this report but modified to express the energy needed as cereal equivalents. Allowances are made for milling and extraction losses, which vary according to the type of cereal and national practices. By specifying these allowances and making use of data on other losses, one can derive a more meaningful assessment of the amount of food aid needed and the urgency of the situation.

The example given in Table 1.2 illustrates the use of the status quo method to calculate Ethiopia's food aid needs and then compares those estimates with a nutrition-based set of calculations.

In addition to estimates of staple food availability, information obtained from surveys of household purchases and home grown food production is used to

Table 1.2 Food availability in Ethiopia, 1987

		Thousand metric tonnes (Cereal Equivalent)
A. Gross production, *cereals and pulses*		
1986/87 Main crop (meher)		6550
1987 Mid-crop (belg)		+ 280
	Total	6830
B. Less seed requirement at 5%		− 340
C. Less post-harvest losses at 15%		− 785
D. Net production, cereals and pulses (*A − B − C*)		5705
E. Enset, roots and tubers		+ 580
F. Milk products		+ 235
G. Domestic production of these four *staple* groups (*D + E + F*)		6520
H. Net annual change in stocks		+ 130
I. *Net* availability from these staples (*G + H*)		6650
J. Assumed 'consumption'* for these stables 1987		7418
K. Deficit of these staple food sources (*J − I*)		768
L. Commercial imports (cereals)		− 200
M. Net deficit (*K − L*)		568

* Estimated from average per annum food supply figures during the base period of 1980/81–1983/4 projected to 1987, and estimated current population numbers.

produce figures for the availability of other non-staple crops. In 1987 these were considered to provide daily an additional 260 kcal per head to the 1480 kcal available in Ethiopia as staples.

Thus a total figure of 1740 kcal was estimated as available for consumption, including 200 000 metric tonnes (MT) from cereal imports. These estimates, based on agricultural statistics and economic data, provide the status quo estimate. On a national basis (expressed in cereal equivalents) the 7418 MT of staples provide 85 per cent of the available energy, other crops contributing 1303 MT per year. In these estimates no adjustments were made for food wastage.

If the different nutrition-based approach is taken, then from the population size, structure, body weights and estimated moderate physical activity patterns, it is possible to generate similar data. In these calculations the infection allowance for children is included but no adjustment to desirable weights for age or height have been made. The comparisons are set out in Table 1.3 and displayed in Fig. 1.13.

Table 1.3 Allowances estimated by the two approaches

	Status quo method		Nutrition based method*	
	M. tonnes × 10⁻³	kcal/head/day	M. tonnes × 10⁻³	kcal/head/day
Staple foods	7418‡	1480	8840 (8039)	1764 (1604)†
Other crops	1303‡	260	1560 (1419)	311 (283)
Total	8721	1740	10 400 (9458)	2075 (1887)

* These estimates include a wastage allowance of 10 per cent as well as the desirable activity and infection allowances of the individuals; the values in brackets exclude the wastage allowance only.

† The partition of energy in the nutrition method between staple foods and other crops is taken to be the same proportion as in the status quo method so that comparisons can be made. Thus 85 per cent of the energy is assumed to be of staple origin.

‡ The tonnage of staples and other crops is converted to cereal equivalents on the basis of their energy contents; 332 kcal/100 g is taken for cereals.

Figure 1.13 illustrates population food needs calculated by the status quo and nutrition-based methods. The food need for the total population is estimated as approximately 20% higher using the latter method. Furthermore, the 'uncovered', i.e. unmet, food deficit rises when using the nutrition-based method by about 30 per cent from 568 and 117 MT (10⁻³) to 1982 and 374 MT (10⁻³) staple and non-staple foods, respectively.

This suggests that either available food production data are inadequate or unreliable, or that the food needs of the population are being gravely under-estimated as a result of using non-biological estimates.

Food deficit regions

Crop failures do not necessarily occur simultaneously in all regions of a country; national food supply figures often conceal a pattern of regional vulnerability. Even if market forces can be relied upon to move food from surplus to deficit

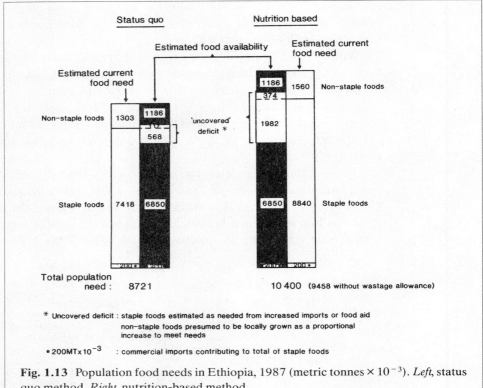

Fig. 1.13 Population food needs in Ethiopia, 1987 (metric tonnes $\times 10^{-3}$). *Left*, status quo method. *Right*, nutrition-based method.

regions, low purchasing power resulting from crop failures will limit the ability of many households to feed themselves adequately. Hence government intervention or external food aid may be needed even when national supplies seem adequate.

1.2 Allowances at individual, household, and national levels

1.2.1 Requirements and allowances for nutrients other than energy

Nutritional requirements, e.g. for vitamins and minerals, are worked out from a knowledge of children's and adults' ability to absorb, store, metabolize and excrete the nutrient. Several figures for the intake of the nutrient may be obtained for each individual, depending on whether one specifies the amount of the nutrient which is required to avoid a deficiency state, to ensure normal body metabolism, or to achieve a certain level of body reserves of the nutrient. Once one has decided on the basis for defining nutrient requirements, it is then necessary to recognize that individuals vary in their requirements, as shown in Fig. 1.14. A dietary allowance, which is usually taken to be the intake needed to cover at least 97 per cent of the population's nutrient requirements, must then be

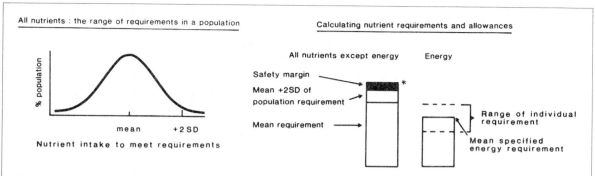

Fig. 1.14 Nutrient requirements and allowances. *Total value is the estimate of the recommended dietary allowance.

calculated. This is usually done by estimating the mean + 2 standard deviations of the observed requirement in a group of individuals. An extra safety margin is often then added to derive the final *recommended daily allowance* (RDA) values as shown on the right side of Fig. 1.14.

This policy is sound because it has been found that if all individuals are provided with the full RDA then those with a relatively low requirement will adapt by reducing the absorption of the nutrient and/or increasing its metabolism or excretion. They do not suffer toxic effects because there is a wide margin of safety in nutrient intakes and the body usually has effective mechanisms for coping with an excess load.

1.2.2 Energy requirements and allowances

Requirements for energy are completely different from the requirements for other nutrients because there is no satisfactory means of adaptation to an energy intake in excess of their individual energy requirements.

If people eat more or less energy than their requirements their body weight will slowly change: any flexibility in their metabolism is surprisingly small and can be neglected for the present purposes of specifying requirements. The relatively fixed nature of each individual's energy requirement and therefore the need to eat a specified amount of food is unique to energy. One cannot therefore recommend an energy *allowance* which is sufficient to cover the range of individual's nutritional requirements and imply that people will actually eat the recommended amount.

If energy needs were estimated by the same method as for RDAs of other nutrients, nearly all the population would exceed their individual requirements, and if they ate their recommended intake they would put on weight with potentially deleterious effects. Therefore energy allowances are not specified by nutritionists. An *average energy requirement* can be listed for a *group* but there is an appreciable range in energy requirements of individuals as shown on the right of Fig. 1.14.

In this manual, however, allowances are given for energy. These are *planning* allowances and not *physiological* allowances. Thus when considering how to provide for the food needs of a population, one is estimating potential additional food needs and energy allowances, should circumstances change. Three types of allowance will be specified in this manual: (a) individual planning allowances; (b) household planning allowances; and (c) national planning allowances.

For simplicity, the word planning will be omitted in subsequent discussions of allowances.

1.2.3 Individual allowances

The different types of individual allowances can be illustrated (Fig. 1.15) by considering 1-year-old underweight children subject to recurrent infections with diarrhoea, temporary lack of appetite and the need to recuperate and to regain body weight.

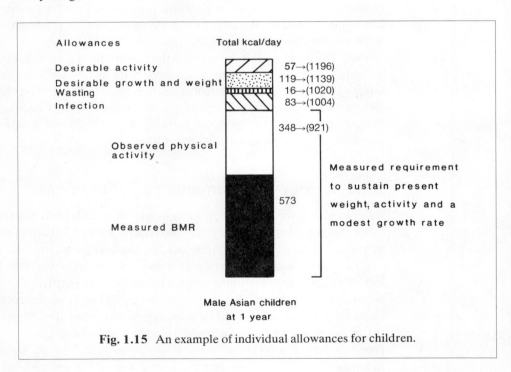

Fig. 1.15 An example of individual allowances for children.

Infection allowances

These need to be applied only to children aged 6 months to 2+ years in less developed countries.

Allowances for wasting

Wasted children are those who are very underweight for their height. However, they form only a small portion of the population in any country unless there are special famine conditions. This allowance for children, specified as the allowance for wasting, is therefore dealt with as a special group problem—see Chapter 8.

Desirable growth and weight allowances

Children in some less developed countries (LDCs) grow more slowly than in affluent societies. This reflects the effect of environmental rather than racial or genetic differences between societies; however, genetic differences may explain some or all of the differences in height between different individuals within a population group. An allowance may therefore be specified to ensure that sufficient food is provided for children to grow at a greater rate and achieve a greater stature in adult life.

Adults are often underweight in LDCs and overweight in developed countries. An allowance can therefore be specified to ensure that adults are able, if need be, to correct their weight so that it is appropriate for their height. However, specifying an appropriate weight for height in many affluent societies will mean defining a weight *below* the average body weight because so many adults in affluent communities are overweight. The effect of choosing a lower weight will be to reduce the energy allowance values for the population. This would be an unusual planning decision.

Desirable activity allowance

This allowance was introduced by the FAO/WHO/UNU 1981 Consultation as a prescription for the better health of the population. In many communities children and adults are too inactive, either because they are limited in their food intake or because they are inactive for cultural reasons.

The impact of these allowances is dealt with in Chapter 6 and summarized in section 1.5.

1.2.4 Household allowance

This may not simply be the sum of the individual's allowances. Two extra allowances may be considered: (a) household distribution allowances and (b) household wastage and storage allowances.

Household distribution allowances

In many societies it is evident that some individuals, e.g. the wage earners, are given preferential access to the particular foods. Some planners have therefore advocated an extra allowance to ensure that any maldistribution of food is counteracted by extra provision. The distribution of food within a household is of crucial importance and although data on this aspect are scarce, the 5th World Food Survey[2] noted that it is common for the diets of children to be 20–30 per cent below their specified energy needs. A bias against females has also been identified in family food sharing in some countries, e.g. Bangladesh and India, but once account is taken of the smaller size and reduced activity patterns of women, the distribution of food seems appropriate to the differing energy needs of the two sexes.

It should be noted that if men are given unlimited access to food, they will only eat according to their needs as determined by their body weights and activity patterns. The additional food will therefore remain for consumption by the other members of the family, or it will be wasted. It is therefore proposed that if full account is taken of individual energy allowances, there is no need to provide an additional distribution allowance to cover the problem of family sharing. In many societies particular care is taken to provide food for the children, but their problems are accentuated by infection. The issue of maldistribution is particularly relevant when a population is malnourished or short of food; men may be given preferential access to food and thus accentuate the problem for the women and children in the household. For the purposes of assessing food needs on a household basis, however, the distribution of food within a family should be assumed to be satisfactory.

Household wastage and storage allowance

Some food will inevitably be wasted except under conditions of meagre supply. In more affluent sections of the community abundant food may be displayed as a sign of wealth. Since the energy requirements of the individuals do not change in response to an excess of available food, this display also implies substantial food waste unless it is stored for later use. Some food is also lost in the household during the processing of grain and other foods. This varies with the type of crop and the degree to which food is processed at local rather than at regional or national level.

Food security at the household level is the fundamental concern of everybody. In most parts of the world some of the food stored at home is lost because it deteriorates and is destroyed by pests, moulds, etc. The effort needed to improve food security within the home may be considerable, since 10–40 per cent or more of a family's total food supply for a year may be lost because of poor food storage. Allowance must be made for this in planning the food supplies of communities who rely on home storage for their needs (see Table 1.4).

1.2.5 National food allowance

In estimating the food needed for a population, care must be taken to allow for the inevitable losses during the storage and distribution of food.

Losses in the food chain may be considerable depending on the effectiveness of the storage conditions before, during, and after the transport of the different commodities, some of which are perishable and liable to deteriorate rapidly.

Many countries make provision on a national, regional, or local basis for seasonal cycles of food availability and ensure that there is enough food stored to feed the population when crops fail or when there are other crises, e.g. floods or hurricanes. The amount of food stored varies depending on government policy

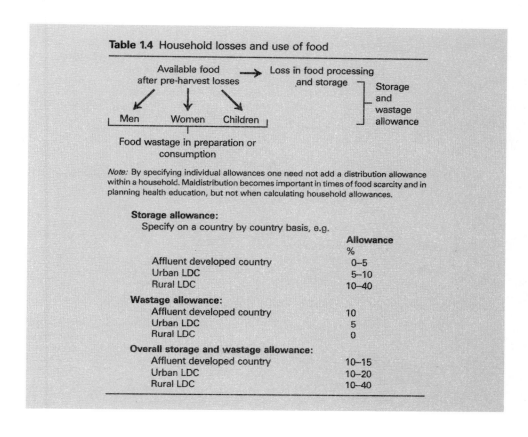

Table 1.4 Household losses and use of food

Note: By specifying individual allowances one need not add a distribution allowance within a household. Maldistribution becomes important in times of food scarcity and in planning health education, but not when calculating household allowances.

Storage allowance:
Specify on a country by country basis, e.g.

	Allowance %
Affluent developed country	0–5
Urban LDC	5–10
Rural LDC	10–40

Wastage allowance:

Affluent developed country	10
Urban LDC	5
Rural LDC	0

Overall storage and wastage allowance:

Affluent developed country	10–15
Urban LDC	10–20
Rural LDC	10–40

as well as food availability. Policy on national food security will be influenced by the nature of seasonal cropping, and the likelihood of crop failure.

There are very few estimates of national storage and distribution losses, but a crude estimation can now be gained by comparing food balance sheet data on food availability with the new estimates of energy requirements produced in this manual. Table 1.5 illustrates the estimates of loss in a country which is predominantly dependent on cassava production. The losses from cassava are small because cassava can be left in the ground in a satisfactory condition until required. Losses from cereal production and the growing of pulses are often greater.

In estimating food supply a range of different estimates is used. The first part of Table 1.5 shows that after allowing for the retention of crops for seed use the following year, and the diverting of food crops for other uses, e.g. animal feed, the food available for consumption in the Ivory Coast amounts to 2445 kcal per day.

The food balance sheet calculations clearly provide estimates of food supply. They can then be used for assessing the adequacy of supply only by making assumptions about the magnitude of allowances at the national, household and individual levels. It becomes apparent that allowances at the individual level are not as important in overall food planning as the household and national allowances relating to food security.

Table 1.5 (a) Food balance sheets: food availability, distribution and losses in Côte-d'Ivoire, 1983–1985. (b) The comparison of estimated individual allowances with food availability to indicate possible losses at each level of the food chain

(a)

	Domestic supply in metric tonnes				
	Production	Imports	Stock changes	Exports	Total balance
Vegetable products	11 049	737	44	1493	10 337
Animal products	205	282	—	38	449

	Domestic utilization in metric tonnes					
	Feed	Seed	Processing	Other uses	Food	Per caput food supply kcal/day
Vegetable products	372	696	1818	2303*	5147	2302
Animal products	—	2	—	8	439	143
					Total:	2445

* This includes the products of alcohol from cassava for fuel use.

(b)

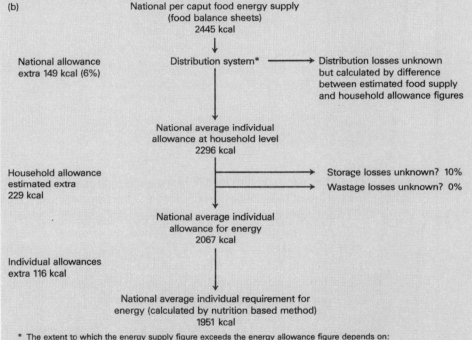

National per caput food energy supply (food balance sheets) 2445 kcal

National allowance extra 149 kcal (6%)

Distribution system* ⟶ Distribution losses unknown but calculated by difference between estimated food supply and household allowance figures

National average individual allowance at household level 2296 kcal

Household allowance estimated extra 229 kcal

Storage losses unknown? 10%
Wastage losses unknown? 0%

National average individual allowance for energy 2067 kcal

Individual allowances extra 116 kcal

National average individual requirement for energy (calculated by nutrition based method) 1951 kcal

* The extent to which the energy supply figure exceeds the energy allowance figure depends on:
1. The accuracy of the figures provided in food balance sheets;
2. The accuracy of the figures used for storage and wastage losses;
3. Distribution losses;
4. The specified individual energy allowances

Note: These household and national allowances are unusually low because the main crop cassava can be left in the ground for prolonged periods with small losses.

The individual planning allowances calculated for the whole of Côte d'Ivoire include an infection allowance for young children and desirable activity allowances in children and adults. To these are added household allowances. By separating each allowance out, Table 1.5 presents estimates of the allowances at each level of analysis.

The magnitude of the projected population growth in the Côte d'Ivoire is of major importance in estimating future food needs. The growth in population size will overwhelm any projected improvements in food security. The relative importance of changes in food security and population growth will vary depending on the country; cassava growing is efficient in terms of storage losses but other countries dependent on cereals or pulses may gain much greater benefit by concentrating on limited storage and distribution losses.

Table 1.5 also serves to illustrate the differences between a status quo approach to food availability and the nutrition-based assessment of food needs described in the section on Food Aid in 1.1. In Table 1.5 the status quo value is 2445 kcal per head per day, whereas the nutrition based method is 1951 kcal per head per day. It is this difference which specifies the magnitude of the various allowances made at a national, regional and household level for losses and other food uses. Unfortunately, in the food aid analysis on Ethiopian data, the status quo value was less than the nutrition-based value. This simply reflects the inadequacy of the figures for food availability and does not challenge the perspective set out in Table 1.5. The status quo figures can be used to obtain some indication of trends in availability, but unless more accurate figures are obtained, one cannot be sure of accurate assessments of the need for food aid, or analyse whether improvements in food distribution can reduce food losses and provide more food in edible form for the population.

1.3 Simplified approach to calculating population energy needs

Traditionally it has been usual to take a single figure for the average energy needs of a population and then to apply this figure to a variety of circumstances. This approach now needs to be modified so that proper allowance is made for the fact that body weight and physical activity are the two prime determinants of energy requirements. Adult groups of different weight will have varying energy needs even if their physical activity patterns and other environmental conditions are the same. Similarly children, adolescents, and adults of different ages and sex also have different energy needs. Therefore different age groups have to be considered separately as tabulated in Table 1.6.

The 1985 report[3] provides a system for establishing:

First, the needs of the body under conditions of absolute rest and when fasting, i.e. under 'basal metabolic' conditions. Simple equations for calculating the basal metabolic rate (BMR) from the body weight of different age and sex groups were derived from a new analysis of the literature[4].

Second, the energy needs of the body for physical activity and the absorption, distribution and storage of food energy are allowed for by multiplying the BMR by a simple value applicable to each group according to their physical activity

Table 1.6 Required aggregations of population data. The choice of age grouping has, for reasons of clarity and to comply with convention[5] been designed as follows: the range, e.g. 7–8, starts at 7 years up to and not including 8 years, i.e. 7–7.99. This range is expressed in tables and figures as 7+, 8+, 9+ etc.

Age years	Population numbers Male	Female
0+	_____	_____
1+	_____	_____
2+	_____	_____
3+	_____	_____
4+	_____	_____
5+	_____	_____
6+	_____	_____
7+	_____	_____
8+	_____	_____
9+	_____	_____
10+	_____	_____
11+	_____	_____
12+	_____	_____
13+	_____	_____
14+	_____	_____
15+	_____	_____
16+	_____	_____
17+	_____	_____
18–29+	_____	_____
30–59+	_____	_____
>60	_____	_____

pattern: the physical activity level (PAL). These values have been derived from measurements of the energy cost of specified activities made in different population groups.

The estimation of the average total daily energy requirement (T) of a group is calculated as: $T = BMR \times PAL$, where BMR is the average basal metabolic rate of the group and the physical activity level (PAL) is the estimated average degree of activity of the group (Fig. 1.16). The basis for considering T as having only two components is given in Chapter 2.

There are technical and practical difficulties which affect the way in which energy requirements are determined. Different approaches have to be used in calculating the energy needs of infants and children (0–9+ years) and of adolescents and adults (10–>60 years). BMR data are available for all age groups, but the additional energy needed to sustain appropriate growth rates and acceptable activity levels in children have not yet been established. The 1985 report therefore follows previous practice for children from birth to 9+ years and relies on estimates of food intake which are expressed as kcal/kg. Energy requirements for adolescents and adults, however, are developed from observations on energy expenditure.

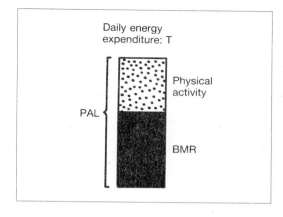

Fig. 1.16 Estimated average total daily energy requirement of a group.

Table 1.7 lists the equations for deriving the BMR from the average weight of each group from 10 years and above.

Table 1.8 provides PALs for the estimated average physical activity of each age group in the whole population. These values have been derived from studies on groups of children and adults in both less developed and developed countries.

Table 1.7 Equations for calculating 'basal' energy needs. These equations are designed for use with the average weights of population groups. These weights can be the actual weights of the population or idealized weights which may be used to allow for normalizing the underweight or overweight population. The equations are extracted from the 1985 FAO/WHO/UNU Report[3]. Details of their preparation are also available[4].

Basal metabolic rate for adolescents and adults

	Males	Females
In kcal per day		
Adolescents		
Age range (years)		
10–17+	$17.5 W^* + 651$	$12.2 W + 746$
Adults		
18–29+	$15.3 W + 679$	$14.7 W + 496$
30–59+	$11.6 W + 879$	$8.7 W + 829$
>60	$13.5 W + 487$	$10.5 W + 596$
In MJoules per day		
Adolescents		
10–17+	$0.0732 W^* + 2.72$	$0.0510 W + 3.12$
Adults		
18–29+	$0.0640 W + 2.84$	$0.0615 W + 2.08$
30–59+	$0.0485 W + 3.67$	$0.0364 W + 3.47$
>60	$0.0565 W + 2.04$	$0.0439 W + 2.49$

* W is the average weight in kg.

Table 1.8 Estimated 'desirable'* allowances for physical activity expressed in terms of PAL values for total populations

Children 0–9+ years	PAL Children's energy needs are based on intake related to body weight	
Adolescents†	**Males**	**Females**
10+ years	1.76	1.65
11+	1.72	1.62
12+	1.69	1.60
13+	1.67	1.58
14+	1.65	1.57
15+	1.62	1.54
16+	1.60	1.52
17+	1.60	1.52
Adults 18–59+ years‡§		
LDC	1.82	1.67
DC	1.66	1.60
Elderly > 60 years†		
In all societies	1.51	1.56

* The word 'desirable' is used to emphasize that the 1985 report contained a series of recommendations aimed at increasing the leisure time activity of children, adolescents and adults in LDCs and DCs. The values are therefore higher than those actually observed and specify activity allowances rather than the observed requirements.

† Source: reference 3.

‡ Average desirable national activity patterns assuming 20 per cent and 70 per cent populations in less developed and developed countries respectively live in an urban environment. This estimate is based on averages derived from U.N. population statistics with an arbitrary classification of member countries into those in the developed and less developed world. The proposed average national values for activity allowances of the urban and rural communities are given in Table 4.1.

§ Estimates based on an analysis of data collated by Ferro-Luzzi[6].

1.3.1 Simplest approach to estimating national energy needs

Energy requirements need to be considered for each sex and in specific age groups as suggested by the 1985 report. These age groups are set out in Table 1.6.

The process for calculating energy needs involves making a *series of decisions* about which data to select; this decision-making process is set out in the form of a 'checklist' in Fig. 1.17.. The *process* for calculating total energy requirements is described in Fig. 1.18 and a practical method of arranging the data needed is illustrated in Table 1.9. An example of the actual calculations for an Asian country is given in Table 1.10.

This decision-making process can range from the simplest form presented here, to more sophisticated analyses based on the availability of additional information which are covered in Chapter 3. Chapter 6, in contrast, deals with the effects of different assumptions on the energy requirement of the population.

These approaches depend on: (a) availability of national data;
(b) the desired complexity of analysis;
(c) whether needs are to be based on current patterns of growth, body size and activity or on 'desirable' targets.

The simplest approach can be organised as a series of decisions made in response to the following questions:

Question 1: Do you have the total population and its distribution by age and sex?

Yes: use own data and proceed

No: then take estimates made for your own country (Appendix 1)

Question 2: Do you have the weights of children and adults?

Yes: use own data and proceed

No: then take estimates made for your own country (Appendix 2)

[The effect of taking optimum growth rates and weights of the population is considered as a separate exercise (Chapter 6)]

Question 3: Do you have data on the basal metabolic rate (BMR) of adolescents (10–17+) and adults?

Yes: use own data and proceed

No: then take standard formulae based on weight from Table 1.7

Question 4: Do you know the national physical activity patterns and energy costs of the adolescents (10–17+) and adults? Note: values for infants and children (0–9+) provided in Table 1.9

Yes: Convert own data on activity patterns to ratios of BMR (see Chapter 3)

No: then assume population activity allowances based on
(a) crude estimates in Table 1.8
or (b) urban/rural population structure (Chapter 4)
or (c) employment statistics (Chapter 3)

Question 5: Do you know how many women are pregnant each year?

Yes: use own data and proceed

No: then take numbers of births (Appendix 1) and calculate number of pregnant women (Chapter 5)

Fig. 1.17 Checklist of information necessary for estimating population energy needs. Proceed to Fig. 1.18 and Table 1.9 for process of calculating total energy allowances.

1. Specify population numbers in each age and sex group (Column *B*).

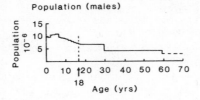

2. Assign average body weights to each age group (Column *C*) (using own data or Appendix 2).

3. For *infants* and *children* (0–9+) calculate individual requirement (Column *G*) by multiplying average body weight (Column *C*) by requirement per kg (Column *F*).

4. For *adolescents* and *adults* (10–>60):

   i. Calculate basal metabolism for each age band (Column *D*) (using equations provided in Table 1.7).

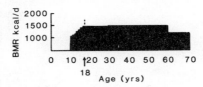

   ii. Select appropriate PAL value (Table 1.8) for LDC or DC (Column *E*)

   iii. Calculate individual requirement in kcal/day (Column *G*) by multiplying BMR (Column *D*) by PAL value (Column *E*).

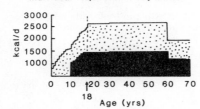

5. Calculate total energy requirements by age group (Column *H*), by multiplying kcal/day by population (*B* X *G*) and thus:

6. Calculate energy requirement of pregnant women in population (*I*) (method in Chapter 5).

7. *Total national requirement* is the sum of group energy requirement (Column *H*), i.e. Paragraphs 3 + 4 + 6 above.

8. *Calculate per caput* needs by dividing *J* (i.e. the sum of Column *H*) by the total population (i.e. the sum of Column *B*).

Fig. 1.18 Process for calculating total energy requirements. Column letters (**B–G**) refer to those in Table 1.9. Calculations are based on age bands specified in Tables 1.6 and 1.9.

Table 1.9 The calculation of total energy allowances in either a Less Developed Country (LDC) or a Developed Country (DC)

A	B Population (thousands)	C Average body wt (kg)	D BMR based on weight (B)	E PAL	F Energy allowance kcal/kg	G Av. individual need total kcal/day (0–9+ yrs C × F) (> 10 yrs D × E)	H Total age group energy need kcal/day × 10⁻⁶ (B × G)
Males				LDC/DC	LDC/DC		
0+			—	—	109/103		
1+			—	—	108/104		
2+			—	—	104		
3+			—	—	99		
4+			—	—	95		
5+			—	—	92		
6+			—	—	88		
7+			—	—	83		
8+			—	—	77		
9+			—	—	72		
10+				1.76	—		
11+				1.72	—		
12+				1.69	—		
13+				1.67	—		
14+				1.65	—		
15+				1.62	—		
16+				1.60	—		
17+				1.60	—		
18–29+				1.82/1.66			
30–59+				1.82/1.66			
>60				1.51			
Females							
0+			—	—	109/103		
1+			—	—	113/108		
2+			—	—	102		
3+			—	—	95		
4+			—	—	92		
5+			—	—	88		
6+			—	—	83		
7+			—	—	76		
8+			—	—	69		
9+			—	—	62		
10+				1.65	—		
11+				1.62	—		
12+				1.60	—		
13+				1.58	—		
14+				1.57	—		
15+				1.54	—		
16+				1.52	—		
17+				1.52	—		
18–29+				1.67/1.60			
30–59+				1.67/1.60			
>60				1.56			

I. Extra energy allowance of pregnant population: _____

Population total: _____   J. Total population energy allowances: _____

$$\text{Per caput allowances} = \frac{\text{Sum of Column H}}{\text{Sum of Column B}}$$

Note: These allowances exclude allowance for desirable growth rates and adult body weights and are based on observed weights of children and adults (see Chapter 5).

Table 1.10 An example of the desirable energy allowance of the population of an Asian country (LDC)

A	B Population (thousands)	C Average body wt (kg)	D BMR based on weight (B)	E PAL	F Energy allowance kcal/kg	G Av. individual need total kcal/day (0–9 + yrs C × F) (> 10 yrs D × E)	H Total age group energy need kcal/day × 10⁻⁶ (B × G)
Males							
0+	10 644	6.6	—	—	109	719	7 657
1+	10 396	9.3	—	—	108	1 004	10 442
2+	10 187	10.9	—	—	104	1 134	11 548
3+	10 014	12.4	—	—	99	1 228	12 293
4+	9 870	13.9	—	—	95	1 321	13 033
5+	9 751	15.5	—	—	92	1 426	13 905
6+	9 651	17.1	—	—	88	1 505	14 523
7+	9 565	18.7	—	—	83	1 552	14 846
8+	9 488	20.2	—	—	77	1 555	14 758
9+	9 414	21.8	—	—	72	1 570	14 776
10+	9 347	23.7	1 066	1.76	—	1 876	17 532
11+	9 289	26.1	1 108	1.72	—	1 905	17 699
12+	9 194	29.3	1 164	1.69	—	1 967	18 082
13+	9 040	33.4	1 236	1.67	—	2 063	18 652
14+	8 847	38.4	1 323	1.65	—	2 183	19 313
15+	8 657	43.4	1 411	1.62	—	2 285	19 781
16+	8 462	47.8	1 488	1.60	—	2 380	20 140
17+	8 268	51.0	1 544	1.60	—	2 470	20 419
18–29+	84 753	51.1	1 461	1.82	—	2 659	225 334
30–59+	111 672	51.1	1 472	1.82	—	2 679	299 125
> 60	26 152	51.1	1 177	1.51	—	1 777	46 473
Females							
0+	10 064	6.1	—	—	109	665	6 692
1+	9 813	8.6	—	—	113	972	9 536
2+	9 596	10.5	—	—	102	1 071	10 277
3+	9 411	12.1	—	—	95	1 150	10 818
4+	9 253	13.4	—	—	92	1 233	12 407
5+	9 117	14.6	—	—	88	1 285	11 714
6+	8 999	15.9	—	—	83	1 320	11 876
7+	8 894	17.4	—	—	76	1 322	11 761
8+	8 798	19.2	—	—	69	1 325	11 656
9+	8 707	21.3	—	—	62	1 321	11 498
10+	8 623	23.8	1 036	1.65	—	1 710	14 745
11+	8 547	26.7	1 072	1.62	—	1 736	14 839
12+	8 444	30.0	1 112	1.60	—	1 779	15 024
13+	8 294	33.5	1 155	1.58	—	1 824	15 132
14+	8 114	36.8	1 195	1.57	—	1 876	15 223
15+	7 937	39.7	1 230	1.54	—	1 895	15 038
16+	7 756	41.6	1 254	1.52	—	1 905	14 778
17+	7 577	42.7	1 267	1.52	—	1 926	14 591
18–29+	77 320	42.9	1 127	1.67	—	1 881	145 475
30–59+	105 319	42.9	1 202	1.67	—	2 008	211 451
> 60	25 683	42.9	1 046	1.56	—	1 632	41 927

Population total: 758 927 (1000s)

I. Extra energy allowance of pregnant population: 4 908 × 10⁻⁶

J. Total population needs (kcal/day × 10⁻⁶) 1 480 697

Per caput requirements = 1 951 kcal/day

Note: Calculations based on actual body weights and FAO (Schofield) equations (1985) using activity values for an LDC and the allowance for infection.

1.4 Comparison of the 1973 and the 1985 recommendations on energy requirements

After the 1973 report[7] had been produced, the FAO secretariat calculated the average national energy requirements, making use of unpublished population statistics and estimates of adult body weights. These figures have been widely used by the UN agencies in assessing food programmes, agricultural development projects, national food supplies, and a variety of other programmes.

A comparison of these figures with those derived from the 1985 report is not easy, but Table 1.11 displays data calculated for several countries using precisely the same criteria as those chosen in 1973, viz. the same adult body weights, the same population numbers and structure, and a uniform moderate activity pattern. The principal difference is that in the 1973 calculations a consistent figure for energy requirement was applied regardless of the observed weights of infants, children and adolescents, whereas in the 1985 calculations (column *B*) we have applied data relevant to 1973 to the weights of all the children and adults. In column *C* we have applied the 1985 calculations to the 1988 data-base using a moderate desirable physical activity allowance and an active pregnancy allowance for all countries.

It is clear that there is in general a trend to decreased values using the 1985 method on the 1973 population data (Table 1.11, column *B*). When requirements are re-estimated by the new method and using the 1988 data-base (column *C*), the requirement figures often increase once more but there is no consistent pattern of change. This is partly a result of differences for individual countries in the weight data and partly the result of changes in the population age structure. In column *C* no changes in physical activity have been assumed. The major

Table 1.11 Average energy requirements kcal/caput/day*

	A†	B†	C
Method:	FAO 1973	FAO 1985	FAO 1985
Data-base:	FAO 1973	FAO 1973	FAO 1988
Country			
Brazil	2174	2077	2107
Burkina Faso	2157	2064	2050
Chile	2220	2096	2142
Ethiopia	2120	1975	1884
India	2006	1957	1925
Italy	2290	2255	2382
Japan	2125	2155	2234
Mexico	2114	2021	2084
Philippines	2051	1886	1925
Thailand	2020	1882	1994
Turkey	2293	2205	2308
USA	2397	2285	2389

* Wastage allowances have *not* been added to any of the values.
† The values in columns *A* and *B* have been calculated using the same population distribution, adult body weight and level of activity.

importance of the 1985 approach, however, is that we now have much greater confidence in the validity of the method, and there is the further possibility of specifying different assumptions and thereby deriving different figures appropriate to the policy of each country.

1.5 Summary of the principal determinants of a population's energy allowance

This manual identifies and quantifies the basis for calculating the energy requirements of a group or of the whole population. Extra increments are specified, including allowances for infection, appropriate child growth, changes in adult stature and weight, etc. This approach therefore differs from the 1973 method of calculating requirements where a variety of allowances were not specified and activities were simply classified as light, moderate, or heavy. The new options allow more specificity but require judgements to be made about the well-being of the population.

The impact of these allowances is illustrated in Fig. 1.19, which sets out on the left the calculated needs of an Asian country based on the assumption that the adults in the population are all engaged in moderate activity. This assumption is the same as that used in generating the 1973 estimates of energy requirements. With the new method of calculating requirements, it is now possible to project the effect of changes of child growth and to specify an allowance which will allow the population to engage in a variety of socially and physically desirable

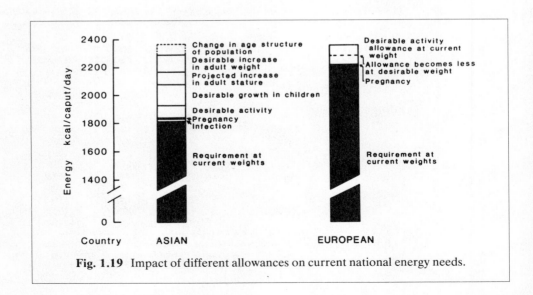

Fig. 1.19 Impact of different allowances on current national energy needs.

activities. Figure 1.19 shows that these allowances make a substantial difference to the estimated needs, with a long-term increase from 1820 kcal/head per day to 2364 kcal per day. This large increase would then bring the energy needs of this Asian country close to that currently needed in those European countries where the population remains active.

The requirement estimates in both countries are based on the same energy values for the metabolism of the children and adults; the effect of climate is negligible. The difference between the requirements of the Asian and European countries is therefore mainly due to the lower body weights in Asia. A small adjustment is also necessary because of the different age structures in the two communities, with the European population having a smaller proportion of children and therefore a higher per caput energy requirement.

The range of options given for the Asian country in Fig. 1.19 are applicable to all countries but particularly those which are less developed; individual national allowances will vary from country to country and can be specified with considerable accuracy providing the age structure, weight and activity patterns of the society are known. The range in energy values to be considered may, however, be much greater than the difference between these two countries. Figure 1.20 and Table 1.12 show the interactions between physical activity and the characteristics of different populations observed in different parts of the world. On the base-line is presented a series of populations *with their appropriate age structures* starting with, on the left, a North American type population which is predominantly old and with high growth rates and high adult body weights. A series of hypothetical countries is then displayed with, on the far right, a less developed country. This population has children with very poor growth rates and the population is becoming increasingly younger and has thin adults. The complete range in energy values is striking and shows the importance of specifying the weights and activity patterns of the populations.

Weight and population structures interact with physical activity; a full range of possibilities has been shown. Thus an activity factor of 1.2 is the starvation allowance suitable for feeding a population group in an emergency where they are *totally* inactive. The 1.4 value signifies a maintenance condition which still leaves no energy for work. At a maximum of 2.0 the population group would be engaged in very hard physical activity—again on an emergency basis. The energy needed in different circumstances can therefore vary 2-fold, with body weight dominating physical activity as the determinant of energy requirements.

This manual can be used to derive simple estimates of energy requirements or complicated analyses of the energy needs under many different circumstances. Nevertheless the factor which overrides the whole of these analyses and adjustments is that of population numbers. The size of the population dominates the demand for food on a national basis.

These new estimates can now be used in planning food policy and in assessing food balance sheets and a variety of economic indices. The new approach also allows a coherent analysis of the components of food loss in the food distribution chain and offers a format for the analyses of household consumption patterns and those of other small groups.

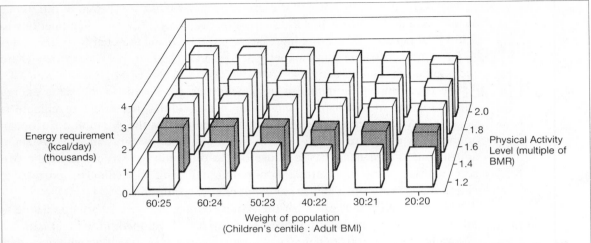

Fig. 1.20 Energy requirements for a hypothetical population of 100 million people: the effects of weight and activity.

Table 1.12 The numerical values for the per caput requirement (kcal/head/day) for a range of body size, physical activity level and population structures shown in Fig. 1.20

Physical activity level	Children's growth pattern, i.e. centile: adult's body mass index, i.e. wt/ht²							
	60:25	60:24	50:23	40:22	30:21	20:20	10:19	3:18
2.0	2949	2900	2828	2651	2588	2384	2350	2203
1.8	2654	2610	2546	2387	2330	2146	2116	1984
1.6	2360	2321	2264	2123	2072	1909	1882	1765
1.4	2066	2032	1982	1859	1815	1672	1648	1545
1.2	1772	1743	1700	1595	1557	1434	1414	1326
Population structure*	Predominantly old			Increasingly older		Predominantly young		Increasingly younger

* These population structures are discussed and illustrated in Chapter 4.

References

[1]*Monthly Bulletin of Economic and Agricultural Statistics* (1973); **22(9)** (Sept.), FAO, Rome.

[2]Food and Agriculture Organisation of the United Nations (1987). *The Fifth World Food Survey*, FAO, Rome.

[3]FAO/WHO/UNU (1985). *Energy and Protein Requirements. Report of a Joint Expert Consultation.* WHO Technical Report Series No. 724, WHO, Geneva.

[4]Schofield, W. N., Schofield, E. C. and James, W. P. T. (1985). Basal metabolic rates: review and prediction. *Human Nutrition: Clinical Nutrition*, **39C**, Suppl. 1: 1–96.

[5]Eveleth, P. B. and Tanner, J. M. (1984). *Worldwide variation in human growth.* International Biological Programme 8, CUP, Cambridge.

[6]Ferro-Luzzi, A. (1984). *National energy requirements of FAO member countries.* Unpublished report to FAO, Rome.

[7]WHO Technical Report Series No. 522. WHO, Geneva, 1973.

2 Principles of energy balance and energy needs

For those not versed in nutrition it is important to emphasize the principles of energy balance in man. Food preparation and variation in food consumption patterns are an everyday occurrence, so it is all too easy to consider that one's own experience provides useful information relating to energy intake, physical activity and weight stability. Unfortunately, research has shown that few assessments by individuals of the amounts of food they consume, or the amount of physical activity taken are reliable. Fluctuations in body weight are also confusing, since they depend in the short term more on changes in water balance than on changes in energy stores. The definition of malnutrition or energy inadequacy may also be confusing, as is the basis for providing a 'normal' amount of energy for a child or adult. The demand for a single value for energy needs applicable to all groups also leads to a great deal of misunderstanding. The purpose of this section is to provide a simple introduction to energy balance and energy need.

The constancy of energy

The standard unit of energy is the joule and human energetics are usually expressed in terms of kjoules (i.e. joules × 1000). A megajoule (MJ) is 1000 kjoules. One kcalorie or Calorie = 4.184 kjoules.

It is a fundamental principle of thermodynamics that energy cannot 'disappear'. Thus (Fig. 2.1) if a group of adult men eat on average 2510 kcal (10.5 MJ) energy daily then that energy has to be either excreted in the faeces, or absorbed by the body. Once absorbed, a small amount of energy is excreted in the urine as the by-product of protein metabolism and the rest of the absorbed fuel has to be metabolized for energy or stored in the tissues as protein, fat, or as carbohydrate in the form of glycogen. The energy derived from the absorbed fuel is used for the multiplicity of chemical processes within the body, and for maintaining the tone of muscles and the movement of the body. If the energy absorbed is not used up by metabolism, then it has to be stored.

Stored energy

Figure 2.1 shows that of the three storage compartments of the body, fat is by far the largest component of stored energy, with an energy reserve in normal Western man equivalent to about 60 times the daily energy need. A temporary fall in energy intake below the rate of energy expenditure is therefore readily dealt with by drawing on energy from the body's stores. A temporary increase in

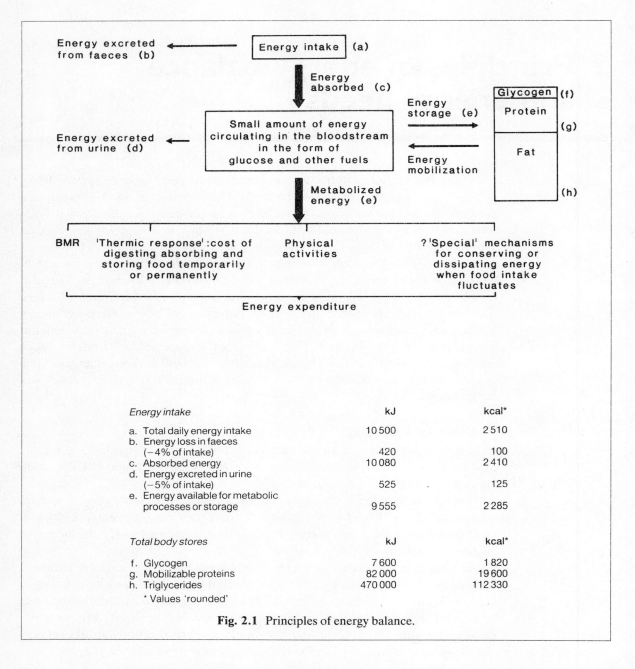

Fig. 2.1 Principles of energy balance.

Energy intake	kJ	kcal*
a. Total daily energy intake	10 500	2 510
b. Energy loss in faeces (−4% of intake)	420	100
c. Absorbed energy	10 080	2 410
d. Energy excreted in urine (−5% of intake)	525	125
e. Energy available for metabolic processes or storage	9 555	2 285

Total body stores	kJ	kcal*
f. Glycogen	7 600	1 820
g. Mobilizable proteins	82 000	19 600
h. Triglycerides	470 000	112 330
* Values 'rounded'		

food intake leads to the deposition of more energy as fat, protein, or to a small extent as glycogen. As the amount of energy stored fluctuates, there will be only slow changes in body weight because the energy density of fat amounts to about 8840 kcal (37 000 kJ) per kg fat, i.e. 1 kg of fat can provide enough energy for perhaps 3–4 days of total energy expenditure during starvation in a woman who continues to be active. But a 60 kg woman also has about 32 kg water in her body, and the amount of water held in the tissues can fluctuate from day to day.

Thus a 2 kg fall in body weight, which can readily occur, may signify *either* no real change in energy balance *or* an actual loss of 17 680 kcal, (74 000 kJ), i.e. equivalent to over a week's energy needs. Changes in the weight of an individual over a few days are therefore a poor guide to alterations in energy balance. Progressive changes in the average weight of an individual or group over a period of weeks can, however, be taken to signify that changes in energy stores have occurred.

Metabolized energy

Figure 2.1 shows that food energy once absorbed can be stored or used for metabolism. This metabolism provides the energy for making new chemical compounds within the body, fuels the muscular activity required to breathe, digest food, and maintain body posture, and also provides the energy for physical activity. Under conditions of rest and when fasting, the basal metabolism reflects predominantly the internal work of the body. The *basal metabolic rate* (BMR) is a major component of the total energy expenditure of the body if expressed on a daily basis, i.e. by estimating how much energy would be used if an individual lay at complete rest in the fasting state for the whole 24-hour period. When food is eaten some energy is used in the digestive, absorptive and storage processes. This energetic response can be termed the '*thermic response*' to food. This component amounts to about 10 per cent of the total daily expenditure. The third component of energy expenditure is that energy required for all types of *physical activity*. These three components are the principal fractions of daily energy expenditure which were considered in the 1985 report and were simplified as shown in Fig. 2.1 so that the thermic response to food was included in the cost of physical activity (see Chapter 3).

It is often proposed that there is a component of metabolism which is flexible and which can respond to the prevailing state of the body's balance of energy and perhaps to cold and to food ingestion. The existence of this '*special*' *mechanism* and its magnitude is much disputed but would present some difficulty in estimating energy needs if it proved to be substantial and sensitive to rapid fluctuations in food intake. In the Fifth World Food Survey[1] mention is made of a theory that the human body may have the ability to adjust the energy cost of the body's metabolism by as much as 30 per cent. If this does occur then it would allow the body to adapt to either a surfeit or a deficit in energy intake without their being any need to change body weight appreciably, to alter energy stores or to adjust the amount of physical activity. This theory has been noted with great interest by economists and planners because it would fundamentally affect the estimate of the energy needs of the population. The theory was invoked in the Fifth World Food Survey to provide a conservative estimate of the proportion of the population which is classified as undernourished. This is a different concept from that relating to planning needs for the population's average energy requirements (see Chapter 3). The theory of fluctuating metabolic requirements is currently being tested but few experts in human energetics would consider the supportive data sufficient to alter the policy on energy requirements set out in the 1985 report. That report, on the basis of current physiological evidence, excluded this mechanism from its analysis on the grounds that it was, if present, of little quantitative importance when applied to the whole population. The significance of the hypothesis will be considered in Chapter 7.

The reproducibility of energy metabolism

Providing individuals are given a *consistent* food intake, remain at the same weight and perform the same pattern of physical activity, then the constancy of total energy expenditure over 24-hour periods is remarkable, i.e. within 2 per cent on repeated measurements. This has been demonstrated by having men and women live in special rooms, called calorimeters, where their energy expenditure can be monitored minute by minute throughout the day and night to an accuracy of 1 per cent. This is true also of the BMR, which fluctuates by no more than 2 per cent in men studied under standardized conditions. In pre-menopausal women there is a small cyclical change in BMR (Fig. 2.2) amounting to a 6 per cent increase from a low point in the mid-follicular phase, at the beginning of the menstrual cycle, to the elevated BMR in the mid-luteal phase which occurs after ovulation. These small fluctuations in young women are only discernible with

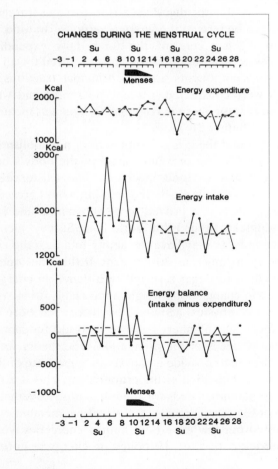

Fig. 2.2 Changes in basal metabolic rate during the menstrual cycle.

accurate methods of measurement and it is often difficult to pick up any differences in the total energy expended by women at different phases of the cycle, even when exercise is standardized, because the subtle effects on basal metabolism are easily obscured.

If energy intake is *changed* on a daily basis by up to 50 per cent of the usual intake, then energy expenditure also changes, but only by about ± 5 per cent. Thus, the effect of short-term fluctuations in intake is to alter that day's energy expenditure by about one-tenth of the change in intake. This response can be thought of as simply reflecting the cost of digesting, absorbing and dealing with the extra food being eaten. It is not true therefore that the body 'burns off' any excess energy that is eaten on a single day—about 90 per cent of it will be stored. If energy intake falls on one day below the average intake, again about 90 per cent of the decrease will be immediately met by the body's own stores of energy. Thus, the body's stores act as a buffer, energy expenditure being finely controlled to meet the needs for voluntary physical activity, and the need for energy to provide the consistent rate of energy used in preserving the normal functions of the body.

Fluctuating energy balance

There is a cycle in energy storage during the day and night. After eating during the day, energy is stored temporarily and then used up while fasting during the night. Spontaneous energy intake fluctuates markedly from day to day, with a coefficient of variation of about 16 per cent. Energy expenditure also fluctuates but to a smaller degree of about 10 per cent, despite the variety of daily activities. Without the fine matching of intake and output, there is often a state of temporary energy imbalance which may last for several days. This occurs normally in a group of people who are, over a period of a month, in a stable energy state. This overall state of energy balance can be illustrated in Fig. 2.2, which shows the daily energy intake and expenditure of an Italian woman over a whole month. As expected, there is a substantial increase in food intake premenstrually with only a small increase in energy expenditure. These changes are almost certainly hormonally induced, so that in this phase of the cycle the individual is in a temporary state where energy is being stored progressively day by day, that is, a state of positive energy balance. Two weeks later food intake is on average less than expenditure, so the individual is in a negative energy balance. Over the whole month, however, the subject is on average in energy balance with no long-term changes in the energy stored as fat, protein or glycogen. Clearly the measurement of a woman's fluctuating food intake is a poor basis for assessing energy needs. In a group of women the *average* intake will be more meaningful, but the requirements for energy by the group really depend on how much energy is being used in the basal metabolic processes and for physical activity.

It is the *longer term average needs* which are of interest to the planner. Energy requirements for planning purposes cannot be linked to the short-term small daily fluctuations in energy expenditure nor to the energy needed to match the demand over a period of a single week. A period of 1–3 months is a more reasonable time span, since over this time an imbalance will slowly become obvious with weight gain or weight loss. To assess energy requirements over a year may be logical for planners, but there will be a need to ensure a reasonable supply of food during the whole year because in some communities there are marked seasonal changes in food supply and physical activity, such as during the wet and dry seasons and when harvesting.

Adjustments to energy imbalance

If there is a food shortage for a period of several weeks, then the individuals within a community will adjust to a deficit in food intake by both drawing on the body reserves of energy and by reducing the components of energy expenditure shown in Fig. 2.1. The basal metabolic rate (BMR) can fall by about 15 per cent after 2–3 weeks of semi-starvation but this will only reduce total energy expenditure by about 8–10 per cent, because the BMR contributes 50–60 per cent to total energy expenditure. This fall in BMR, which has been extensively documented, reflects a metabolic adaptation in adults and children when they are fed on an energy intake which is substantially reduced, e.g. by about 50 per cent. Whether more modest reductions can achieve similar effects on BMR in the long term is unknown.

With the loss of energy from body fat during semi-starvation there is also some loss of body protein. Since this component is metabolically active, there will also be a further decrease in the cost of maintaining this reduced protein mass. Thus, the BMR declines further. When people are semi-starved for 6 months and given only about half their original energy intake, then the fall in body weight is profound. There is a fall in the BMR of about 40 per cent, 15 per cent of this reflecting the early metabolic adjustment, and the other 25 per cent reflecting the progressive loss of metabolically active body protein. This loss of protein is a very frequent, if not inevitable, accompaniment of the loss of body fat.

The thermic response to food will also diminish on semi-starvation because there is less to eat: this may conserve another 2–5 per cent of energy expended but has not been extensively studied.

The two remaining components of energy expenditure shown in Fig. 2.1 are physical activity and the 'special' adaptive mechanism. The 1985 report assumed, on the basis of a literature analysis, that the important adjustable component is physical activity. Thus, the report specified that there are in essence only two principal responses to a fall in food supply which can be documented by simple observation and measurement: (a) *a loss in body weight* as stored body energy is used to maintain metabolism and activity; and (b) *a behavioural change* in physical activity to reduce the amount of work or leisure activity.

More sophisticated measurements allow an assessment of the metabolic adjustments in BMR but these adjustments have only been demonstrated with substantial degrees of energy imbalance. These adjustments are not therefore considered to be the usual response of healthy people living under normal circumstances, but a pathological response: a fall in BMR has only been documented during the process of weight loss. A reduced BMR has not been found in groups who can be classified as well fed, healthy and physically active.

The implications of this new approach in the 1985 report are profound because it is no longer sensible to state, for example, that a weight-stable population has an energy intake of 80 per cent of its energy requirements unless it is recognized that this must mean that energy expenditure is also at 80 per cent of the expected value. Thus, either the population is underweight, or it is very inactive. In theory a population could be poorly nourished and have a reduced BMR before becoming appreciably underweight. Under these circumstances, however, the average weight of the population could be expected to be falling.

These conclusions form the basis of the 1985 report; further analysis of the problem of adaptation is given in Chapter 7.

The principle of assessing energy expenditure rather than intake

The 1985 report, in contrast to that of 1973, emphasized the importance of considering the *energy expenditure* of a population when estimating their requirements for energy. An analysis of food intake data is unhelpful, since as already noted it is possible to have a grossly inadequate food intake in a group of subjects who have responded by weight loss so that they have become substantially underweight. They may only be able to maintain this weight by reducing their physical activity to a minimum, and by reducing the metabolic activity of their body tissues in an attempt to adapt to the low intake. The assessment of energy expenditure is therefore a more logical approach where one can specify the requirement value as being that energy needed for specific components of energy output, such as work and leisure time activity. This does, however, mean that one also needs to specify the appropriate weight of the population and the amount of activity that is considered 'desirable' for the citizens of a particular country. This clearly involves judging the occupational and social needs of the population.

Reference

[1]Food and Agriculture Organisation of the United Nations (1987). *The Fifth World Food Survey,* FAO, Rome.

3 Different levels of analysis in estimating requirements

Only recently has it become possible to measure the total energy expenditure of individuals over an appreciable period of time by the use of a non-radioactive isotopically labelled water method. For many decades nutritionists have relied on meticulous analyses of individuals' metabolism under conditions of rest, eating, and activity throughout a 24-hour period. It is these measurements which have been used for developing an understanding of the energy needs of children and adults.

If one attempts to deal with the energy requirements of the population by detailed analyses of the energy used by a spectrum of individuals engaged in a huge variety of activities, the whole exercise becomes impossibly complicated.

Three major simplifying processes have been developed to overcome these difficulties. They are:

Simplification of energetics. A new approach to specifying the total energy used by people of any age and size relates all the energy costs to their basal metabolic rate.

Simplification of timing of activities. Three stages have been developed to provide detailed, integrated hourly or overall daily activity patterns.

Simplification of activity description. This simplifies the process of describing different activity patterns by using, for example, occupational descriptions. These can be used in terms of hourly or daily activity patterns.

3.1 Simplifying the components of energy expenditure

Realistic attempts to estimate energy requirements demand some simplification of the analysis of energy costs as shown in Figs 3.1 and 3.2. The figures show how the traditional physiological approach to estimating energy expenditure has been simplified by considering only two components rather than three. The basal metabolic rate is measured under somewhat artificial conditions, but it is a long-standing, well-tried and reproducible measurement supported by an extensive literature of data from men and women of different countries, and of different sizes and ages studied under a variety of nutritional conditions. When the size of the individual, i.e. his/her weight and height, is taken into account, then the coefficient of variation between individuals amounts to only 6 per cent for men

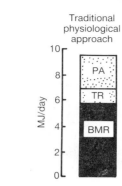

Traditional physiological approach

Physical Activity (PA): this depends on both the occupation and the type and extent of other socially related activities.

The Thermic Response to food ingestion (TR): i.e. the use of energy in digesting, absorbing, storing and disposing of the ingested nutrients.

Basal Metabolic Rate (BMR): i.e. the rate of energy expenditure under standardized conditions of immobility in the fasting state with an environmental temperature of 26–30°C to ensure no activation of heat generating processes.

Fig. 3.1 Principal components of energy expenditure: traditional approach.

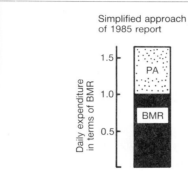

Simplified approach of 1985 report

Physical activities: costs considered to:
(1) include thermic responses to food
(2) be expressed as a multiple of the basal metabolic rate.

Basal Metabolic Rate

Fig. 3.2 Principal components of energy expenditure: simplified approach.

and 8 per cent for women. There is evidence that Indian men and women have basal metabolic rates which are about 9 per cent below those of North Americans or Northern Europeans, but the reasons for this are not known. A reduced BMR is found particularly in undernourished Indians, so the relevance of these findings to 'healthy' people is unclear. Furthermore, there may be a small decline in BMR in relation to environmental temperature, since studies on Europeans in India show that their BMRs tend to fall, whereas Indians living in Europe tend to have BMR values similar to those of Europeans. In the absence of clear data from many countries on the impact of environmental temperature, and recognizing the wide variation of temperature in countries such as India, which ranges from the Himalayas in the north to the tropical south, the 1985 report considered the evidence for any genetic, racial or pervasive environmental factor too slim to warrant the use of any special BMR figures for particular countries. Therefore, unless the planner has a specific need to obtain and use local data on BMR, it is advisable for the present to derive the values as set out in Table 1.7.

Instead of providing data on each of these three components, the 1985 report assessed energy needs by first specifying the needs for covering the basal metabolic rate and then considering all physical activity costs as effectively occurring in the fed state as shown in Fig. 3.2.

The BMR value is the key to the expression of energy requirements because all other energy costs are now considered as multiples of the BMR. Only a small error is associated with including the energy involved in the biological handling of food, i.e. the 'Thermic Response' to food, in the values for physical activity. The dietary component varies a little with the type of food eaten, but is predictable for a group of subjects. To have specified it separately in the 1985 report would have been an unnecessary complication in the procedure for calculating requirements.

3.2 Progressive simplification of estimates of physical activity and its costs

Figure 3.3 provides an example of the energy a group of women might expend over a whole 24-hour period as they lie in bed at night, rise in the morning to wash and dress, then eat; for this example we can propose that they walk to work, where they spend 8 hours, with a 1-hour lunch break, before returning home. There they undertake household chores and meal preparation before going out for the evening to play games or dance for an hour. Obviously the activity pattern, if measured accurately on 100 women of similar age, size and general behaviour, would show a variety of patterns, but the graph shows what could well prove to be minute-by-minute expenditure. If, in addition to such detailed monitoring of individual activity, the cost of each of the physical activities is measured, then one can be confident of having an accurate measure of the true total energy expenditure of this group throughout the day and even during the night. Obviously there will be a range in the normal BMR and in the energy cost involved in different individuals undertaking the same task. The variation in energy expenditure, however, is not large and amounts to a coefficient of variation of about 10 per cent for individuals of the same size. The variation takes account of the small subtle differences in activity as well as the differences in metabolism of individuals within the group. For the purpose of population estimates, however, the average figure will suffice. No country has representative data collected in this way and covering all sectors of society. Indeed, to collect such data would be a very long and expensive task. Some simplifying methods or assumptions must therefore be used.

This discussion of how to simplify the calculation of the energy cost of physical activity deals first with the observed activity patterns of people in a community. Chapter 6 presents an analysis of the impact of desirable changes in body height, weight and physical activity; should one wish to specify desirable allowances, these are detailed in Chapter 4.

Actual energy expenditure over 24 hours

Take 25-year-old women of average weight 50 kg engaged in moderate physical activity at work, e.g. working in a field and involved in housework and social activities, e.g. communal meetings and dances.

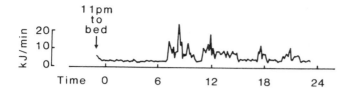

This tracing could be specified if the subjects either occupied a huge measurement chamber with simulated activities for the period or had all their activities monitored with measurements of the energy cost of each activity.

Simplifying approaches to estimating energy expenditure:

Stage 1 (a) Measure times involved in all activities.
(b) Take literature values for estimating energy cost of activities.
(c) Calculate minute-by-minute energy expenditure during day and night.

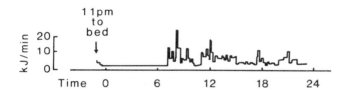

PAR
Physical Activity Ratio = the energy cost of individual activities maintained on a minute-by-minute basis and expressed as a ratio of the BMR

Stage 2 Specifying Integrated Energy Indices after classifying daily energy expenditure into Household (H), Cardiovascular maintenance (C), Occupational (O), Residual time (R), Discretionary activities (D) and time sleeping in bed as equivalent to BMR.

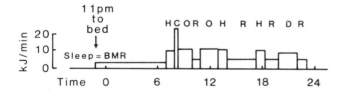

IEI
Integrated Energy Index = the energy cost of a task or during a period of time, including rest periods, expressed as an integrated value for the whole period and in terms of the BMR

Stage 3 Classifying groups simply as of light, moderate or heavy activity.

Moderate activity women
Therefore total expenditure over the day averages 1.64 x BMR

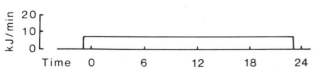

PAL
Physical Activity Level = the 24-h energy expenditure expressed as a ratio of the BMR

Fig. 3.3 Progressive simplification of energy expenditure in the 1985 report.

3.3 Simplification: Stage I: Detailed activity monitoring

Because analyses of the energy cost of different activities is so time consuming, the first stage of simplification (Stage I—Fig. 3.3) dispenses with the continuous monitoring of energy expenditure. If the time allocated to different activities is known, then the actual costs of these activities can be taken from tables which summarize estimates of the energy cost of each activity. In doing this there are two main problems to be taken into account:

1. The method of measuring the time spent on each activity by individuals or groups (*activity pattern analysis*);
2. The method of expressing the actual *energy cost of an activity.*

3.3.1 Activity pattern analysis

When describing individual activities it is important to be aware that the energy cost of similar activities will vary from country to country and will be described differently. For example, in some countries the weeding of a field might involve a great deal of bending with a high level of energy expenditure but in other countries, depending on the crop and level of mechanization, the amount of energy which has to be expended may be much less. Accurate description of activities is important if the resulting energy values are to be used cross-culturally, and if they are to be appropriately classified into group categories (Stage 2).

Detailed analysis

Before the daily energy expenditure can be calculated it is necessary to have information on the amount of time spent on *each* activity, i.e. *activity pattern analysis.* Since the need for the information was recognized, there have been a number of studies to provide data on the activity patterns of different groups, but unfortunately this information is rarely published in a form which allows comparisons between societies. The information can be obtained by observers who monitor individuals minute by minute throughout the waking hours. Self-reporting can also be used but this may be unreliable, particularly in societies where concepts of time and the allocation of tasks on a minute-by-minute basis are different from those, for example, in Western cultures. In some groups, such as the young and old, there may be appreciable differences between the time allocated by self-reporting and that monitored by observers.

Simplified approach

Detailed activity monitoring is an exhaustive exercise which may be simplified by specifying the activities performed in each 15-minute period. This could, for example, be based on the self-reporting of the time spent on each activity throughout the day. Recording the approximate time in hours spent on specific tasks is an even simpler approach. As the monitoring becomes cruder it is more difficult to be certain of the validity of the estimations of energy expenditure because few people at work are continuously active, and the number of pauses during an activity, e.g. in gardening or doing factory work, can make a substantial difference to the final energy cost of the work.

One example of the analysis using a very simplified approach is given in Table 3.1, which shows the proportion of time spent by elderly Italian women and men on different activities. The social environment may make an appreciable difference to the activity patterns, and suitable energy values for these activities have then to be chosen from those shown in Appendix 4.

Table 3.1 Crude analysis of activity patterns of elderly Italian women and men

	Women			Men		
	Retirement homes %	Own homes Urban %	Rural %	Retirement homes %	Own homes Urban %	Rural %
Sleeping	33	32	35	33	33	37
Resting in bed	14	9	8	13	8	7
Sitting	35	33	32	34	38	34
(inactive)	(25)	(25)	(24)	(24)	(31)	(27)
(active)	(10)	(8)	(8)	(10)	(7)	(7)
Standing	3	8	7	3	7	7
(inactive)	(3)	(4)	(3)	(3)	(6)	(5)
(active)	(1)	(4)	(4)	(1)	(1)	(2)
Walking (stairs)	3	2	2	6	5	4
Moving about	11	16	16	10	9	11
Other	1	1	1	1	1	1
	100	100	100	100	100	100

Rounded values.
Example taken from Ferro-Luzzi[1].

3.3.2 Energy cost of activities

Traditionally the energy cost of an activity has been adjusted to take account of the average weight of the group. This was necessary because the total amount of energy used each minute is greater in large than in small people. The energy needed to move about will also be greater in bigger individuals. The 1985 report chose, however, to simplify the calculations by expressing the energy cost of different occupations as physical activity ratios expressed in terms of the estimated BMR rather than the weight of the group. Summary tables of the energy cost of a variety of activities or physical activity ratios have been provided in Appendix 4.

Physical activity ratio (PAR)

The physical activity ratio expresses the energy cost of an individual activity per minute as a ratio of cost of BMR per minute. Detailed calculations show that the cost of women and men undertaking a specific task is very similar for the two sexes once it is expressed as a PAR. The approach also has the merit that people with very different body weights, and therefore very different rates of total energy expenditure, have the same activity ratio. Thus the energy cost of walking on the level at 1–2 km/h for an 80 kg man is 2.0 kcal/min (12 kJ/min) compared with a cost of 2.5 kcal/min (10 kJ/min) for a 50 kg man. When expressed in activity ratios, however, the cost in both cases is similar and amounts to more than twice the BMR. Thus, by calculating the BMR first, we can assign an energy cost ratio of, for example, walking, sewing or hoeing in the garden to groups of either sex and of different weights.

The way in which these individual activities can be incorporated into the overall analysis of the 24-hour energy expenditure is illustrated in Fig. 3.4, which provides an example of the various activities of an adult woman working on agricultural tasks in an LDC. Clearly her day is crowded with a variety of tasks, many of which have different PARs and all of which have to be collated to obtain a reasonable estimate of the overall average rate of her energy expenditure.

The 1985 report did provide preliminary values for the energy cost of each occupation suitable for use on a minute-by-minute basis. Since the presentation of preliminary data in the report was all that was possible, there has been a subsequent attempt to include more of the literature on physical activity costs, where these have been measured with well standardized apparatus. A new set of tables has therefore been constructed which can serve to provide detailed information on energy costs; these are presented in Appendix 4.

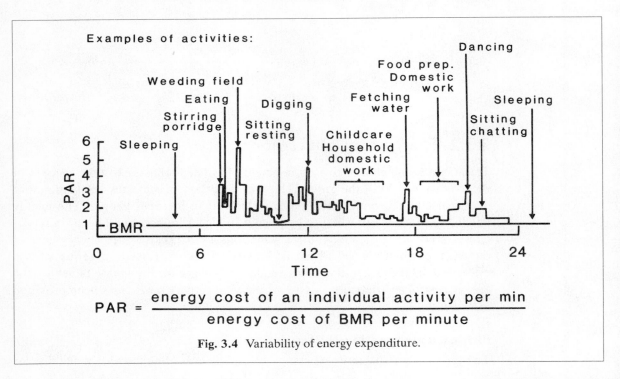

Fig. 3.4 Variability of energy expenditure.

3.4 Simplification: Stage II: Average activity estimates for periods of the day

This second stage of simplification is *probably the most detailed method which government planners might use* in calculating their estimates of requirements. Experts in work physiology and nutrition may well require the more detailed analysis shown in Stage I, but such an analysis on a population basis would prove very time-consuming. With the Stage II simplification, the same problems of activity pattern analysis (3.3.1) and the energy cost of activities (3.3.2) are taken into account and then data are collated for the whole 24-hour period to provide the estimated average daily rate of energy expenditure.

3.4.1 Activity pattern analysis

In the second stage of simplification (Stage II, Fig. 3.3), instead of assessing each individual activity, the activities are grouped into four categories which can be built up to cover the full 24-hour period. The four categories are:

1. Time spent in bed: the overall energy cost is taken to be the same as the BMR;

2. Occupational time (O): energy cost estimates are chosen appropriate to the task;

3. Household tasks (H), non-occupational and discretionary activities (C,D): energy cost estimates are chosen for the overall type of activity;

4. Residual time (R), when individuals are not engaged in major energy consuming activities: the energy cost is designated as 1.4 times the BMR.

These categories are discussed below.

Time spent in bed

For the time spent in bed it is assumed that the energy expenditure is equivalent to the BMR, i.e. the physical activity ratio is 1.0. In practice the rate of energy use during sleep is about 95 per cent of the BMR but this discrepancy has a negligible effect on the estimates of requirement.

Occupational time

This stage of simplification makes an allowance for the pauses in work so that integrated values can be used for different occupations. The values therefore take account of the behavioural patterns of people under varied living conditions. Men and women may undertake a particular task in different ways and at different speeds.

Household tasks and non-occupational or discretionary activities used in developing activity allowances

The non-occupational activities of men and women may be more difficult to assess, but energy is also needed for housework and a variety of leisure time activities.

There are four distinctions to be made:

1. It is recognized that in most societies women undertake more housework than men. The analysis set out in the 1985 report reflects a simplification of the assessment of population allowances by deriving single approximate energy values for a variety of household tasks and for a number of different social activities. Thus, moderately active men were considered to have the need to do some household chores at an energy cost equivalent to that of walking. In addition, they were expected to be involved with other tasks about the house, or in the community, for a further hour at an average energy cost of 3.3 times the BMR. The values chosen differ from group to group but were based on an analysis of what a healthy group in a community might be expected to do.

2. Women in less developed countries without so much mechanization in the home spend appreciably longer over food preparation. A total of 3 hours is therefore estimated as the average time spent on this activity.

3. Men who are either retired or unemployed should, in practice, be allowed an additional hour of discretionary activity.

4. The addition of time for more intensive exercise to maintain cardiovascular tone and muscular capacity was specified in the 1985 report as suitable for all light activity in adults aged 18–59 +, the intensity of the exercise being reduced in the elderly.

Residual time

This is simply the time remaining when children and adults are not engaged either at work or in household and other discretionary tasks, and when they are not resting or sleeping in bed. For this time the 1985 report simply allocated an overall energy value of 1.4 times the BMR.

3.4.2 Energy cost of activities

Each period of activity can now be given an *Integrated Energy Index* which is the energy cost of the activity or occupation including the pauses for rest expressed on a minute or hourly basis and calculated again as a ratio of the BMR. This is very different from the PAR, which is concerned with minute by minute changes in activity and deals with pauses separately.

Table 3.2 extracts from the tables in the 1985 report the Integrated Energy Indices and time allocated for these additional but important activities.

Table 3.2 Integrated Energy Indices of household tasks and other discretionary activities of adults

	Integrated energy index for the whole as a multiple of BMR	Time allocated hours
(a) Household tasks		
Men	2.7	1
Women (D.C.)	2.7	2
(L.D.C.)	2.7	3
(b) Discretionary activities		
(i) Socially desirable:		
Men 18–59+	3.3	1
Elderly Men > 60	3.3	2
Women (D.C.)	3.3	2
Women (L.D.C.)	2.5	2
(ii) Exercise to maintain cardiovascular and muscular function in light activity adults only		
Men 18–59+	6.0	1/3
Women 18–59+	6.0	1/3
Elderly Men > 60	4.0	1/3
Elderly Women > 60	4.0	1/3

Note: These values are taken from the examples listed in the 1985 report in which such indices are used to illustrate the process of calculation, rather than as specific proposals for general application. Nevertheless, in the absence of new and more reliable information, these values may be helpful in deriving crude estimates of the average energy allowances of a community.

Some economic surveys are conducted which attempt to assess the amount of time children and adults spend in specific tasks. If this information is available (as for Stage I simplification) then it is possible to apply the PAR values taken from Appendix 4 to the data. This appendix has been compiled from original data and in this case, the values for both sexes have been combined since it was observed that the energy cost of so many of the tasks was little different once these costs have been converted into values which were ratios of the BMR. This set of values is more comprehensive than that given in Appendix 5 and in the 1985 report, and the values can be used as alternatives when they describe specific occupational tasks. These new values do, however, require that the rest periods are well documented as explained below.

Allowing for pauses during activities: developing Integrated Energy Indices

One can refine the analysis by assessing the amount of time spent in pauses at work, and thereby derive an integrated energy value known as the *Integrated Energy Index*. The term *Integrated Energy Index* (IEI) relates to the average energy cost over the whole period of time allocated to a task. It therefore includes the time taken for pauses between the specific activity which characterizes work: e.g. pauses between cutting down trees; household tasks, e.g. pauses between washing kitchen utensils, and discretionary activities, e.g. pauses between episodes of dancing. It must be emphasized that these Indices are all expressed as a ratio of BMR but the values are lower than the physical activity ratios: the ratios are the actual costs of performing the tasks without any pauses (see footnote *).

Collating the data for the whole 24-hour period

The number of hours worked per day must be specified and an allowance made for the additional discretionary activities. The energy indices covering the integrated energy cost of the work hours are needed, not the PAR values. It is then possible to devise a set of tables as shown in Table 3.3, which provides the simplest tabular form in which to do the calculations.

For calculating the average time spent at work, an adjustment must be made for the fact that many people work for 5 or 6 days per week. The daily average value will need to be estimated accordingly. The energy demands on workers undertaking specific jobs, such as farming, mining, ship-building or tree felling may vary enormously depending on the degree of mechanization, so energy indices appropriate to the degree of mechanization need to be considered.

Table 3.4 provides an example of the calculations for a group of subsistence farmers living in a less developed country. The process is one of simply calculating the BMR of the group from their average weight, i.e. 65 kcal per hour, and then applying this value to the hours spent in bed. Then the Integrated Energy Index for the work is used to multiply the estimated BMR, i.e. $3 \times 65 = 195$ kcal per hour. This value is then multiplied by the total average time spent on the work to give the total value of 1365 kcal per day. Then the energy cost of an hour's household tasks are added, and a further hour is allowed for discretionary

* The relationship between the PAR and IEI in the two sexes.
There appears to be a conflict between the observation that the energy costs of an activity is the same in men and women of whatever body weight (Appendix 4) and the finding (see below) that women have daily energy levels expressed in terms of their BMR which are lower than those of men. The difference is small for light activities but more marked for moderate and heavy activities. The explanation for the different Integrated Energy Indices in men and women is that preliminary analyses of women's working patterns suggest that they have longer pauses than men, so that the cumulative energy cost of the task or the whole of the time spent working is less. This was taken into account in the 1985 report when generating overall values for a day's energy requirement. In the 1985 report the absence of reliable data on the time spent at rest made it difficult to produce useful data on Integrated Energy Indices. The approach was therefore to assign equal rest periods to men and women but to give men rather higher PAR values when they were engaged in the specific tasks, so that the overall work output over periods of several hours corresponded with the known collated data on the energy costs incurred over these time periods. It is these integrated values which are presented as Integrated Energy Indices in Appendix 5. When good data are available on the time spent at work and at rest, reliance should be made on the PARs shown in Appendix 4 and then new Integrated Energy Indices appropriate to the pattern of work in the occupational group under study should be produced.

The crude estimates of the Integrated Energy Indices serve to emphasize how little is known of the activity pattern of different groups in many societies. If more sophisticated analyses are to be used, then clearly there is a need for detailed studies of the frequency and duration of rest periods if reasonably accurate estimates of energy expenditure are to be obtained.

In some societies workers may continue their task with barely a pause, e.g. when workers are working with conveyor belts in automated factories or when women are weaving steadily for hours on end. If this applies, then it is possible to further refine the analysis of energy costs by taking the PAR values shown in Appendix 4 and then applying different lengths of time to the rest or pause periods.

Table 3.3 Estimating desirable activity allowances for each population group. (First calculate BMR per day from Table 1.7 and divide by 24 to provide BMR values per hour)

	Time taken hours	Ratio Energy to BMR (IEI)	Energy need for each task
1. Sleep and bed rest	e.g. 7	1.0	_____
2. Work activity	e.g. 6	Appendix 5	_____
3. Household tasks	e.g. 1–2	Appendix 5	_____
4. Socially desirable activities	e.g. 2	Table 3.2	_____
5. Cardiovascular maintenance	e.g. 0–3	Table 3.2	_____
6. Residual time	e.g. 7–8	1.4	_____
		Sum:	_____
		Average energy allowance:	_____

Note: In this table it is necessary to choose the appropriate amount of time spent on several activities. This choice will affect the final value for the energy requirement.

Table 3.4 An example of the steps needed in calculating the desirable activity allowances of subsistence farmers. Males aged 18–29+ years, average weight 58 kg. Estimated basal metabolic rate from equations in Table 1.7: 65 kcal per hour

Activity	Time hours	Integrated Energy Index	Total energy cost* kcals
In bed	8	1.0	520
Occupational activities	7	3.0	1365
Household tasks at 2.7 × BMR	1	2.7	176
Other discretionary activities	1	3.3	215
For residual time needs	7	1.4	637
Total daily energy allowance (rounded):			2910

The 24-h energy allowance (2910 kcal) expressed as a ratio to BMR (1560 kcal), i.e. the physical activity level (PAL) = 1.87

* *Note:*
(a) *Calculating total energy allowance from hourly data*
  1. Estimate BMR for body weight using predictive equations (Table 1.7);
  2. Estimate BMR per hour by dividing BMR value by 24;
  3. Calculate the energy cost of each activity per hour by multiplying the Integrated Energy Index for that activity by the BMR expressed in kcal; and
  4. Multiply this total energy cost for each activity by the hours spent on each activity;
  5. Sum the different activity expenditures to give the total daily energy requirement, allowing all the residual day time to have an Integrated Energy Index of 1.4.
(b) *Calculating PAL value*
  Take the final total energy requirement figure and divide by the BMR value obtained in Stage I of the calculation.

activities at $3.3 \times$ BMR before the remaining time is specified as residual time at $1.4 \times$ BMR. The total energy then comes to 2910 kcal per day.

Table 3.4 is an example of calculating activity allowances for a particular occupation. The approach opens up the possibility of building up a set of values for each work group in the population using simple observations on their activity patterns and the type of job. Statistical data on occupational groups and some very simple observations on how each group spends their leisure time may therefore provide reasonable estimates of each group's energy expenditure. This further option allows the planner to build up relatively easily a composite picture of the whole society.

3.5 Simplification: Stage III: Single values for the whole 24 hours

Clearly it should be possible to simplify the estimation of energy needs even further if the household tasks and desirable social activities can be included in a general estimate of the whole day's energy needs based on occupation. This is illustrated at the bottom of Fig. 3.3, where a single value is taken to reflect the overall pattern of desirable physical activity of the group of women, and expressed once more as a ratio of BMR.

At this stage of simplification it is possible to select a single physical activity level (PAL) value for male and female adults, set at light, moderate or heavy activity levels (Table 3.5).

3.5.1 PALs for male and female adults

The values for these activities, expressed on a daily basis, are termed the PAL values to distinguish them from the physical activity values (PAR). The PAR refers to the minute-by-minute estimation of the energy cost of the specific physical activity itself, which is in turn different from the Integrated Energy Index. The Indices are the integrated values, including the pauses, for the time taken to perform tasks during the day. The overall integrated values for the PAL for men and women engaged in light, moderate, and heavy physical activity are given in Table 3.5, which simply shows how the men have a slightly lower Physical Activity Level at the light activities, but a higher PAL at the moderate and heavy activities. It is these which were used in developing the simplest approach to the calculation of population energy needs.

Table 3.5 Average daily desirable activity allowances of adults expressed as a PAL value

	Type of work (desirable activity allowance)		
	Light	Moderate	Heavy
Men	1.55	1.78	2.10
Women	1.56	1.64	1.82

Desirable activity values taken from 1985 FAO/WHO/UNU Report[2].

3.5.2 Simplification of activity description

This is the third type of simplification set out in this chapter and depends essentially on defining occupational groups. Incorporated within the collation of time into hourly periods is a simplification of physical activity in the house into a general category of household tasks. Leisure time activity has also been given the general description of socially desirable activities. These generalizations helped in the collation of time and will not be considered further.

3.5.3 Use of employment statistics in occupational analysis

This approach has been developed in this manner because many countries have their own unpublished statistics on their economically active population which can be used. Another source of data is the International Labour Organisation, which produces an annual *Yearbook of Labour Statistics*[3]. The data are arranged as far as possible in accordance with standard classifications such as the International Classification of All Economic Activities (ISIC) and others.

An approach which depends on labour statistics is hampered by a number of problems:

1. Variations occur between and within countries in the method of data collection and tabulation. Differences in the definition of occupations and groups covered also mean that data are not always comparable.

2. International statistics define only those who are economically active and who furnish the supply of labour 'for the production of economic goods and services as defined by the UN systems of national accounts and balances' during a specified period of time.

3. In general, such data sets do *not* include the young, the elderly, students, dependents and women occupied 'solely in domestic duties'. The activity rates for women are therefore considerably underestimated because in many countries they engage in agricultural and other family enterprises without pay, and therefore do not qualify for classification.

Nevertheless, it is possible to obtain estimates of the actual energy cost of different groups of workers by classifying the different occupations listed in the International Labour Organisation reports. The suggested values can be listed as in Table 3.6 to provide composite values for the Energy Indices, assumptions having been made about not only the variety of tasks undertaken by each category of worker, but also the intensity of the work and the amount of time allowed for short rest periods while working.

3.5.4 PAL for occupational categories

Rather than dealing with light, moderate and heavy activity levels, the planners

Table 3.6 Classification of occupational activity as Integrated Energy Indices according to occupational categories listed by the International Labour Organisation[3]

Occupational group	Activity classification	Occupational Energy Index for time spent at work	
		Male	Female
Professional, technical and related workers	Light	1.7	1.7
Administrative and managerial	Light	1.7	1.7
Clerical and related workers	Light	1.7	1.7
Sales workers	Moderate	2.7	2.2
Service workers	Moderate	2.7	2.2
Agricultural, animal husbandry, forestry, fishing and hunting	Mixed– Mod./Heavy	3.0	2.3
Production and related transport equipment operators and labourers	Mixed– Mod./Heavy	3.0	2.3

Note: These values are estimates and may need to be revised in the light of futher work.

may wish to estimate the energy needs of the population according to occupational category. In theory, the classification of the international labour organization can be extended to provide single figures for the total energy need of the group. An example of how to calculate the PAL of a subsistence farmer from composite values is shown in Table 3.4, where the PAL value is 1.87. Suggested PAL values expressed on a daily basis for different occupational groups are displayed in Table 3.7. An approach which uses occupational categories is dependent on the availability of appropriately disaggregated population data.

Table 3.7 provides some guidance by taking, as noted earlier, the work categories listed by the International Labour Organisation. On this occasion the groups are amplified to take account of the elderly, the unemployed, domestic workers and women as well as men who are self-employed, e.g. in subsistence farming, and therefore not generating the income required for classification in the ILO scheme.

3.5.5 Mechanization: potential effects on activity patterns

The mechanization of work and of transport and the use of mechanical aids in the home has had a profound effect on the pattern of activities over the last 50 years in many affluent developed countries. It is therefore to be expected that a classification of occupational activities in accordance with the ILO categories shown in Table 3.7 can provide only some indication of the range of activity levels; these will differ markedly depending on the degree of mechanization. Thus, even professional and clerical workers in less developed countries do more in the home, in travelling to work and when at work than in a highly mechanized industrialized society.

Mechanization makes a great difference to the amount of energy which has to be expended in a given task, so the demands on the agricultural worker in a less developed country are substantially different from those made on a mechanized

Table 3.7 The development of the International Labour Organisation's classification to derive simple estimates of PAL values for each occupational group separated into averages for developed (DC) and less developed (LDC) countries

Occupational group	Occupational PAL values			
	Males		Females	
	DC	LDC	DC	LDC
1. Professional, technical and related workers	1.55	1.61	1.56	1.58
2. Administrative and managerial	1.55	1.61	1.56	1.58
3. Clerical and related workers	1.55	1.61	1.56	1.58
4. Sales workers	1.67	1.78	1.60	1.64
5. Service workers	1.67	1.78	1.60	1.64
6. Agricultural, animal husbandry, forestry, fishing and hunting	1.78	1.86	—	1.69
7. Production and related transport equipment operators and labourers	1.78	1.86	1.64*	1.69
8. Housewives	—	—	1.56	1.64
9. Students	1.55	1.61	1.56	1.58
10. Unemployed	1.55	1.61	1.56	1.58
11. Subsistence farmers	1.78	1.86	1.64	1.69
12. Domestic helpers	—	1.78	1.60	1.64
13. Elderly > 65	1.51	1.51	1.56	1.56

* Excludes female labourers.

Note: These values are based on different proportions of the population being assumed to be of light, medium and heavy activities. The assumption is made that in less developed countries there will be a greater demand for physical activity because the mechanisation of transport, work, home and leisure time activities will be appreciably less than in developed countries.

farm in an affluent society. In some countries forestry or transport is very mechanized, in others highly seasonal; in some the demands could correctly be considered appropriate to light activity, in others to heavy. It is difficult to overcome these objections without much more detailed knowledge of the work pattern and job intensity of different groups of workers.

References

[1]Ferro-Luzzi, A. (1987). *Time allocation and activity patterns of the elderly.* Background paper prepared for FAO, Nat. Inst. Nutrition, Rome.

[2]FAO/WHO/UNU (1985). *Energy and protein requirements. Report of a Joint Expert Consultation.* WHO Technical Report Series No. 724, WHO, Geneva.

[3]International Labour Organisation (1986). *Year Book of Labour Statistics.* I.L.O., Geneva.

4 Impact of urbanization and population structure

It is common for economic analyses to consider urban and rural areas individually, so a separate analysis of energy requirements of the two populations is also sometimes needed.

To obtain these requirement values the following data are ideally required, as would be expected from the simplified scheme for estimating energy needs set out in 1.3.

1. Population numbers in urban and rural areas categorized by sex and into the age groups set out in Table 1.6.
2. Weights and heights of all the age and sex groups in the two areas;
3. Activity patterns of all the groups in the two areas;
4. Total annual number of pregnancies in each area.

Clearly these data may be not available in full, so it is necessary to make assumptions and to choose approximate values.

4.1 Data in urban and rural populations

Many countries have their own unpublished information on the distribution of populations in urban and rural areas. The United Nations does have some information, but there are several disadvantages with the routine application of this information to planning:

1. The data are recognized to be of variable quality.
2. Much of the information is out of date and not comparable from country to country. Thus the tables available at an international level do not apply to the same period of time and may on occasion be 2 decades out of date.
3. The UN aggregation of data into 5-year age bands does not allow their use in the new scheme for calculating energy needs. Thus recalculating the information into appropriate age groups would first be necessary.

Given these problems, data available within each country are likely to be much more useful.

If data are unavailable, one may simply assume that the age groups are evenly distributed between the rural and urban areas and that the two populations' weights are similar. One can then apply estimates of activity patterns to the two populations, these being based on a number of surveys.

4.2 Activity patterns

The rapid and progressive increase in the size of urban communitites in many countries leads to a variety of social and environmental changes which may be difficult to assess in terms of energy requirements. It seems likely, however, that the pattern of occupational and leisure time activities in urban societies involves less energy expenditure than in a rural environment. Unfortunately no studies are available which systematically allow an analysis of the changes in activity patterns which occur as children and adults migrate from a rural to an urban area. On the basis of a variety of data it is suggested that the activity of those in urban and rural communities can be classified as shown in Table 4.1.

It is known from demographic data that the distribution of the population by age and sex between the urban and rural areas is not uniform. FAO analyses[1] suggest that young and elderly women tend to remain in rural areas to a greater extent than men. Table 4.2 illustrates this, and shows the surprising point that

Table 4.1 PAL values for use in urban and rural communities

		Both Developed (DC) and Less Developed (LDC) countries			
		Urban	Rural	Urban	Rural
Adolescents*	Age	Male/Female		Male/Female	
	10	1.76 1.65		1.76 1.65	
	11	1.72 1.62		1.72 1.62	
	12	1.69 1.60		1.69 1.60	
	13	1.67 1.58		1.67 1.58	
	14	1.65 1.57		1.65 1.57	
	15	1.62 1.54		1.62 1.54	
	16	1.60 1.52		1.60 1.52	
	17	1.60 1.52		1.60 1.52	
Adults					
Males		1.61†	1.78‡	1.67§	1.86¶
Females		1.58	1.64	1.60	1.69
Elderly*					
Males		1.51**		1.51**	
Females		1.56		1.56	

Note: The PAL values are taken from estimates derived by Ferro-Luzzi[1] from analyses of limited published and unpublished data.

  * Children, adolescents and elderly are accorded equivalent activity patterns in urban and rural communities in both developed and less developed countries. This approach is therefore a normative one and does not reflect actual practice. In the absence of any comprehensive analysis this approach seems most appropriate, as suggested by Ferro-Luzzi.[1]

  † For adults in urban DCs assume 75 per cent light and 25 per cent moderate activity levels.

  ‡ For adults in rural DCs assume 100 per cent moderate activity level.

  § For adults in urban LDCs assume 50 per cent light and 50 per cent moderate activity levels.

  ¶ For adults in rural LDCs assume 75 per cent moderate and 25 per cent heavy activity levels.

  ** For elderly (> 60 years) in LDC and DCs for both urban and rural groups assume 100 per cent light activity levels.

Table 4.2 Categories of urbanisation and the age groups showing the greatest differences from the average[2]

Urban category	% Population living in urban areas	Age groups (y) % in urban areas			
		0–5 y Males	25–49 y Females	> 64 y Males	> 64 y Females
1	12.3	12.0	8.6	9.7	11.2
2	30.0	30.5	24.5	27.6	28.2
3	44.6	52.3	42.5	48.0	46.4
4	70.0	73.1	64.9	70.7	67.9
5	84.4	86.5	82.9	88.6	81.1

boys under 5 years of age tend to be more numerous in urban communities despite the proportion of young women being below average.

Figure 4.1 reproduces a map of countries classified using the crude urbanization figures available to UN sources. The low urbanization levels characteristic of Category 1 are mostly found in the Sahel, East Africa, the poorest African countries, and in some parts of Asia. Category 2 or urbanization includes the rest of the Central and Southern African region. In Category 3 nearly half the population live in towns as in Eastern Europe, USSR, North Africa, and in countries in the Near East. With nearly three-quarters of the population living in the towns in Category 4, one is dealing with several of the most developed countries in the Northern Hemisphere but countries which still retain an appreciable rural community. The final category, with very small rural populations, comprises Australasia, a few European countries, and some from Latin America.

In order to illustrate the effects of the different activity patterns associated with urbanization alone, i.e. without considering the differences in sex distribution displayed in Table 4.2, a calculation has been made assuming that all age groups are equally distributed between urban and rural communities but that the proportion of each age group in the towns is 12.3, 44.6 or 84.4 per cent. On this basis the energy allowances are, on average, 2123, 2094 and 2055 kcal/head/day, respectively, on a national basis. It was also assumed in these calculations that the average weight of each group was the same in the urban and rural areas throughout the urbanization process. Thus, what is observed is simply the effect of the urbanization process itself in a community which had a structure with a predominantly young population. If this West African country also developed rapidly and became affluent, with more mechanization at work and in the home, then the activity patterns of the urban people would be expected to fall further and simulate those observed in a developed country. The choice of activity patterns in urban and rural areas makes a modest impact on the overall calculation of energy requirements. To obtain accurate values it is necessary to refine one's understanding of what the population's activity patterns really are in the two areas.

To illustrate this approach, Table 4.3 shows recent data analysed by FAO on the estimated PAL values calculated from a very extensive survey of activity

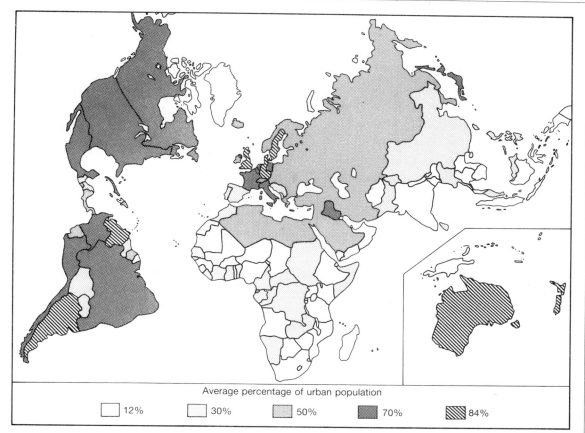

| | 12% | | 30% | | 50% | | 70% | | 84% |

Fig. 4.1 Countries classified according to the crude urbanization figures available to UN sources.

patterns in Finland. The table shows that when the survey was conducted in 1979, the adolescents living in towns tended to take more, not less exercise than the adolescents living in the rural areas. The PAL values for the two adolescent populations were, however, close to the estimates of adolescent allowances suggested in the 1985 report. Adult males tended to be more active in towns and less active in the country than the estimated average allowances for developed countries provided by the 1985 report, but the estimates shown in Table 4.1 seemed appropriate for both urban and rural women.

An analysis of the prevailing activities, stature and age distributions of urban and rural populations, conducted on a preliminary basis by FAO,[2] suggested that the urbanization values seemed, on a country-by-country basis, to be dominated by the fact that those countries with a high degree of urbanization tend to have heavier children and adults and an age structure with a higher proportion of adults. This is not unexpected if it is recognized that countries with well established urban communities are usually affluent with a well grown population and a low mortality rate. These associations on a cross-country basis will minimize the impact of lighter activities on energy requirements induced by

Table 4.3 The estimated PAL values for the physical activity of a large random sample of children and adults in Finland[2]

Age group	Urban population		Rural population	
	Males	Females	Males	Females
10–11	1.68 (1.74)*	1.55 (1.64)	1.65 (1.74)	1.54 (1.64)
12–14	1.65 (1.67)	1.63 (1.58)	1.64 (1.67)	1.61 (1.58)
15–17	1.62 (1.62)	1.56 (1.53)	1.59 (1.61)	1.53 (1.53)
18–19	1.65 (1.61)	1.57 (1.58)	1.65 (1.78)	1.57 (1.64)
20–24	1.67 (1.61)	1.58 (1.58)	1.69 (1.78)	1.59 (1.64)
25–29	1.67 (1.61)	1.60 (1.58)	1.71 (1.78)	1.60 (1.64)
30–39	1.68 (1.61)	1.61 (1.58)	1.73 (1.78)	1.64 (1.64)
40–49	1.69 (1.61)	1.62 (1.58)	1.72 (1.78)	1.65 (1.64)
50–59	1.65 (1.61)	1.62 (1.58)	1.72 (1.78)	1.63 (1.64)
60–64	1.56 (1.51)	1.60 (1.56)	1.62 (1.51)	1.59 (1.56)

* Proposed national averages from Tables 1.8 and 4.1 developed from the 1985 report are inserted in brackets for comparison.

urbanization. However, the planner who seeks to assess the projected energy allowances as urbanization progresses has to make separate judgements and not base them on cross-cultural comparisons.

It may be possible to develop estimates of energy allowances based on the occupational analyses and suggested work indices presented in Appendix 5 and Table 3.6. By using these values planners may specify the average number of hours worked per week (calculated on an annual basis to take account of festivals and holidays). They can then build up the average day of their population in urban and rural areas as though dealing with separate nations. Thus the average energy requirements of males and females of each group can be calculated in each area before the two population groups are collated for producing an average national figure.

4.3 Population structure and changes

The proportion of young and old people within a population varies enormously from country to country. There are also rapid and progressive changes occurring dependent on the relative birth and age-specific mortality rates. Much of what the planner needs to know will depend on projecting estimates for the population size and structure over a long period of time. This manual is not the place for a detailed analysis of the process by which these estimates are derived and it is assumed that this process is familiar to readers. This manual will concentrate on illustrating firstly the effects of having different age structures within a society and secondly what the effects of a projected agricultural change might have on predicted future energy needs.

From an analysis of the population structures in 142 countries, it has been proposed that the age pyramids of different countries can be classified approximately into four types (Fig. 4.2), the proportions for children and adults in the four model countries being shown in Table 4.4. The first type depicts the pattern found for example in Scandinavia, Northern Europe, and North America, where there is a substantial proportion of elderly people; the second population pattern, i.e. with a population growing increasingly older, applies for example to Central European countries. The third model, with a predominantly young population, reflects the impact of a high birth rate and the greater life

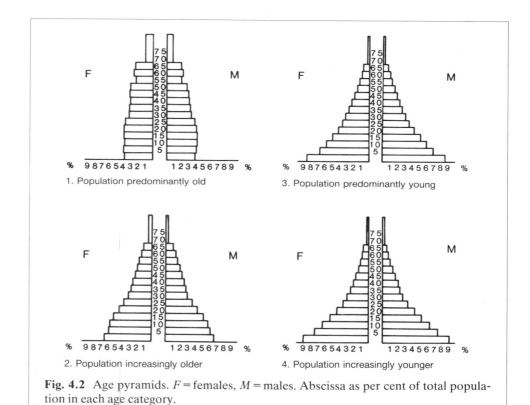

Fig. 4.2 Age pyramids. *F* = females, *M* = males. Abscissa as per cent of total population in each age category.

Table 4.4 The proportion of the total population in each of four types of age pyramid depicting different patterns of national population structure[1]

Type of population	% Total population in each age group		
Age group (y)	0–19	20–64	> 64
1. Predominantly old	32.3	56.1	11.6
2. Increasingly older	47.1	47.7	5.2
3. Predominantly young	53.0	43.8	3.2
4. Increasingly younger	56.5	40.6	2.9

These average values for the four types of population were derived from multiple correlation analyses of national population statistics.

expectancy now being observed in many less developed countries such as in Asia and South America. The fourth type depicts a population which is growing younger, a type which is less common than the other three, but nevertheless represents the age pyramid found in some African countries.

One way of illustrating the effects of this different age structure is to assume that a hypothetical African population corresponding to type 4 goes through progressive changes whereby it eventually attains the characteristics shown for a predominantly old population (Type 1). Table 4.4 shows the proportions in each age group for the four types of population structure. To illustrate only the effect of different age structures on the average energy allowance, the total population numbers are assumed to remain constant (100×10^{-6}). (The impact of population change will be given below.) The calculations also assume that there are no changes in the activity patterns of a country as it changes its population structure. This is again an unlikely assumption, the effect of which will be reconsidered later. Figure 4.3 shows the total energy allowances at each age for each population model.

This display of the total population energy needs in each age group shows how

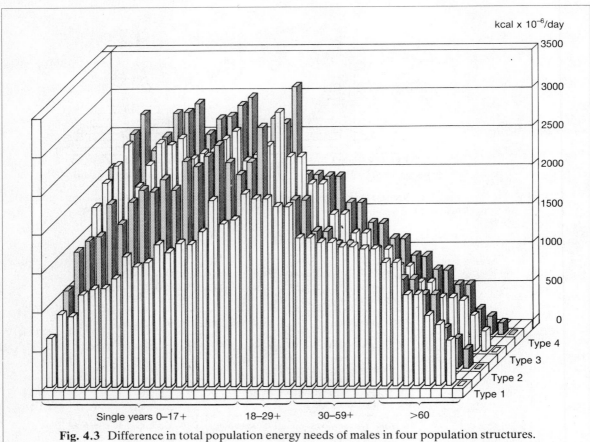

Fig. 4.3 Difference in total population energy needs of males in four population structures.

the needs of the young are emphasized in the Type 3 and Type 4 population structures. Since the scale of the axis is different for children and adults it would be wrong to conclude that the children's energy needs dominate national energy requirements.

Table 4.5 sets out both the total and average energy allowances to be expected for the hypothetical 100 million population shown in Fig. 4.3. Although the differences are clearly discernible and alter by about 7 per cent, the effects would be overwhelmed were the population to increase in total by 20 per cent i.e. from 100 to 120 million people. The 7 per cent change depending on age structure, compares with the urbanization effects of about 3 per cent. Weight changes would have greater impact (see Chapter 5) but it is the value for the total population which dominates the calculation of total national energy requirements.

Table 4.5 The average and total energy allowances of hypothetical populations of 100 million people with four possible types of population structure set out in Fig. 4.2

Type of Population	Energy allowance	
	Average kcal/head/day	Total kcal × 10⁻⁶
1. Predominantly old	2203	219 806
2. Increasingly older	2144	212 069
3. Predominantly young	2092	210 237
4. Increasingly younger	2068	205 836

Note that for Fig. 4.3 and Table 4.5 all adults are assumed to be moderately active, with an infection allowance for young children; child and adult weights are those observed in the Côte d'Ivoire, and the BMR equations are those proposed in the 1985 Report. (The birth rate is held constant at 340 per 1000 women in the age group 15–29 + years and an active pregnancy allowance is assumed for these calculations only). All these assumptions remain the same in the present calculations so that the difference observed in the table reflects only the effects of different population pyramids. In Fig. 4.3 each column reflects the total energy allowance in kcal × 10⁻⁶ for each single year group in childhood. For adults each column no longer represents a single year of age, but a series of columns corresponds to an adult age band. In order to accommodate the limited space 18–29 + year olds are represented by only 5 columns, the 3 decades from 30 to 59 + years by 8 columns and for the over 60s three columns represent the progressive decline in total energy needs of the age group up to an arbitrary limit of 75 years. The dominant difference between the types is in the distribution of total energy needs in different age groups, which is uneven as would be expected from Fig. 4.2.

References

[1]Ferro-Luzzi, A. (1984). *National energy requirements of FAO member countries.* Unpublished report to FAO, Rome.

[2]Bério, A. J., François, P., and Périssé, J. (1985). *Population dynamics, anthropometry and urbanisation; their relationship to human environment.* Paper for the International Union of Scientific Study of Population (IUSSP), 5–12 June, Florence, Italy. In Food and Nutrition, **11**, (1), 1985 (FAO).

5 Energy allowances

5.1 Introduction

The 1985 report specified energy need as 'that level of energy intake which will balance energy expenditure when the individual has a body size and composition, and level of physical activity, consistent with long-term good health, and which will allow for the maintenance of economically necessary and socially desirable physical activity'.

The 1985 report established new criteria for estimating energy requirements by highlighting the distinction between a population's actual energy expenditure and their energy needs based on a specific set of 'desirable' standards. We have therefore highlighted the differences by specifying that requirement figures are different from energy allowances. Requirements relate to healthy people undertaking their normal life without being subject to energy restrictions.

In assessing the energy needs of young children within a population in a less developed country, the 1985 report provided estimates of the energy needed by apparently healthy but underweight children to grow at a rate which would *not* allow for catch-up growth. Similarly, the specified energy need did *not* take account of the intermittent periods when food intake falls because of recurrent infections. This is because the policy of the Consultation was first to specify only the energy requirements of healthy children and to deal with special allowances for sick or convalescing children separately.

For adults the energy used in occupational activities and obtaining food is clearly fundamental and can therefore be considered as a true requirement. For long-term health the individual or community may, however, require additional food. It is these increments identified in the 1985 report which are designated in this manual as allowances. These extra allowances are summarized in Table 5.1.

These allowances are added to the observed energy requirements of children and of adults who are of inappropriate weight or who do not have a level of activity considered as 'desirable' for social or physical reasons. A further allowance is made for young children subject to recurrent infections. These allowances may be achieved either by increasing the level of intake per kg of the individuals' actual weight, or by maintaining the energy value per kg but presuming that the increased weight has already been achieved, thereby effectively overestimating their actual requirements.

In order to make clear the nature of these allowances, each will be considered separately before their additional effects are examined.

Table 5.1 Summary table. The choice of energy values and body weights for specifying extra allowances in children and adults

Type of allowance	Group affected	Childhood energy values	Data sources	Weight	Comments
			Adolescent and adult physical activity level pregnancy allowance		
1. Requirement	All groups	Requirement Table 5.8	Requirement level Table 5.8	Actual	Maintains status quo
2. Infection allowance	Infants 6 months–2+ years	Infection allowance Table 5.7	—	Actual	Applied to LDCs; incorporates desirable activity allowance
3. Wasting allowance	Infants and children, adolescents	Infection allowance Table 5.7	Requirement level Table 5.8	Desirable *weight for actual height* 0–9+ yrs. NCHS in Appendix 3.1 10–17+ yrs. Baldwin in Appendix 3.2	Applicable to conditions of severe deprivation or famine
4. Desirable weight allowance	Adults	—	Requirement level Table 5.8	Desirable *weight for height* Fogarty in Appendix 3.4	Optimum for normalizing weight in both LDCs (underweight: often in combination with wasting allowance for < 18) and DCs (overweight)
5. Desirable growth allowance	Infants and children, adolescents	Requirement values Table 5.8	Requirement Table 5.8	Desirable *weight for age* 0–17+ yrs. NCHS in Appendix 3.5	Normalizes for stunting and stunting and wasting. Optimum choice for LDC 0–17+ years
6. Desirable physical activity allowance	All groups	Desirable allowance Table 5.8	Desirable level Table 5.8	Actual	Proposed in 1985 report all LDC and DC countries for use when no adjustment in weight or height needed

Note: Energy needs, other than for the maintenance of status quo [1] are specified in one of two ways:
either   (a) by multiplying the actual weight by an increased intake expressed: in childhood as kcal/kg act. wt. [2, 6]
                or   in adolescents and adults as an increased PAL [6]
or   (b) by calculating with an increased weight as if the population has already achieved the expected size, and multiplying by the requirement values for childhood intake or adult PAL [3, 4, 5].

5.2 Allowances for the different growth patterns of children

Many populations have children who grow relatively slowly because of a combination of repeated infections and an inadequate food intake. Intake may fall by 20–40 per cent during an infection, and further losses of body energy and protein occur because of vomiting, diarrhoea, and the metabolic losses induced by fever. The 1985 report accepted that children of different ethnic groups have the same growth potential and that the causes of widespread growth retardation are principally environmental.

Children can be underweight for their age for two reasons: (i) *stunting*, i.e. a failure to grow in height, and (ii) *wasting*, i.e. being too thin and underweight for height. Clearly children can be both stunted and wasted, in which case they will have an even greater deficit in weight for their age.

Stunted children have failed to grow in height, but they are usually of normal body proportions and therefore have a normal weight for their reduced height. If one therefore takes the current weights of children for calculating energy needs, this automatically means that in many LDCs one is dealing with groups of the population who are 'underweight' for their age. Catch-up growth in height, to recover from stunting, is possible but occurs at a slower rate than catch-up growth in weight for height (wasting). This second cause of underweight is unusual and seen particularly during emergencies such as famines. These children have an increased risk of morbidity and mortality. This type of severe weight deficit is generally found in only 2–10 per cent of children under the age of 5 years.

Three forms of requirement allowance based on a child's weight can be specified.

1. *Energy requirement.* This specifies the energy need of a child based on its existing height and weight even if the child is stunted and/or wasted.

2. *Wasting allowance.* This allowance is for those groups of children who have an average weight below the 50th percentile weight for their actual height. As noted above, this condition is unusual except under conditions of severe deprivation. The allowance is calculated by applying the same energy requirement figure in kcal/kg to an increased weight, i.e. equivalent to the desirable weight for healthy children of the same height. To normalize for wasted children's underweight for height, the NCHS 50th centile weight for actual height values should be taken from Appendix 3.1. In addition, children less than the age of 3 years will also be allowed an infection allowance; this should be provided for all young children in less developed countries.

3. *'Desirable' growth allowance.* This allowance is for children who have not grown as well as healthy children and are stunted for their age. This stunting then means that they are also underweight for their age but of normal body proportions for their reduced height. Stunting in relation to the WHO reference values based on the North American growth standards is extremely

common in LDCs. The allowance is made by assuming that children are up to the standard weight for age when applying the kcal per kg energy requirement value. Thus the children are provided with an energy intake well in excess of their immediate requirements, the extra allowance then providing the opportunity for catch-up growth. The 50th percentile values for weight for age of healthy children to be used in these calculations are given in Appendix 3.5.

The distinctions between these approaches to energy allowances and their effects on the estimated energy intake of a single group of 18-month-old girls is shown in Table 5.2. In this table it is assumed that all but child A are stunted and that they may also be wasted. The allowance for infection, to be considered next, is also included for convenience. Table 5.2 provides four allowance values for the energy needs of children.

It is evident that an allowance for underweight children can be substantial. Thus the wasting allowance increased the allowance of these children from 876 kcal/day to 989 kcal per day. This would provide an extra 113 kcal/day which is sufficient for the weight deficit of 1.1 kg to be corrected over a period of several months, this time varying depending on the proportion of lean and fat tissue laid down.

If an extra allowance is made for desirable growth, i.e. to combat the stunting,

Table 5.2 The effect of different weights in calculating energy requirements and allowances for 18-month-old girls who live in a less developed country and who are either healthy and growing well, or displaying various degrees of undernutrition

	Child	A Healthy and well grown	B Stunted	C Stunted and wasted	D Stunted and wasted with recurrent acute weight loss
Height	Centile	50th	3rd	3rd	3rd
	cm	80.9	75.1	75.1	75.1
Weight for height	Centile	50th	50th	10th	10th
	kg	10.8	9.6	8.5	8.5
Energy *requirement* for child's actual weight kcal per kg		103	103 (108)*	103	108*
Total energy *requirement* kcal/day		1112	989 (1037)*	876	918*

* *Infection allowance:*   i.e. rapid weight loss with infection imposes the need to replete the weight deficit.

Wasting allowance: this provides child C with the allowance of child B.
Desirable growth allowance: this provides children B, C and D with the allowance of child A.

Note: All these figures, whether allowances or requirements, exclude a 5 per cent extra component for desirable physical activity which will be discussed separately. Only two options are worth considering in planning in less developed countries. The first is to include the infection allowance. Thus, in the usual example of a stunted but non-wasted child B, a weight value of 9.6 kg can be used with an energy allowance figure of 108 kcal/kg, i.e. a total of 1037 kcal/day. To this value the 1985 report routinely added 5 per cent for desirable physical activity, thus providing 113 kcal/kg and a total intake of 1085 kcal. The second option is to provide the desirable growth allowance, i.e. 1112 kcal which assumes that the children are at a normal weight for age, i.e. child A. This extra energy provides for complete normalization of the height deficit and could lead to a more sustained catch-up in height; it increases the child's requirement by a further 7 per cent. To this can be added the 5 per cent for desirable physical activity, bringing the total to 1166 kcal for child A.

then the children would be considered as though equivalent to child A and therefore to have an energy need of 1122 kcal/day. This amounts to an extra 236 kcal for the wasted and stunted child C. This would allow the child to grow from the lower centiles for both weight and height back to a presumed 50th centile value, designated as the median value for a healthy population, i.e. like child A. This is the desirable growth allowance.

If children were only provided with food in amounts based on their low weight for age, this would still allow them to continue to grow along their reduced centile but would not provide enough food for the progressive catch-up and maintenance of a higher weight for age.

A decision to provide children with an energy allowance for desirable growth results in a substantial increase in the total energy allowances of young children as illustrated in Table 5.2. It also affects the average allowance for an LDC childhood population as shown in Table 5.3. The increase in energy allowances was 9 per cent if allowances were related only to children's desirable weight for actual height, but correcting also for the height deficit, i.e. providing a desirable growth allowance, leads to an increase of over 25 per cent for the children below the age of 10 years.

The impact of these allowances on the overall population figure can be considered in a somewhat artificial way by assuming that only children less than 10 have the allowances applied to them. In this example (Table 5.4) no infection allowances are included, as this is dealt with separately (see below).

Table 5.3 The effect of different weight allowances for children below 10 years of age on the energy allowances (kcal/head/day)

	Requirement only	Wasting allowance only	Desirable growth allowance only
LDC	1157	1276	1489
DC	1431	1434	1488
Criteria used:	Actual weight	Desirable weight for actual height requirement with no infection allowance	NCHS weight for age requirement

Table 5.4 The impact on a population's energy requirement (in kcal/head/day) of specifying different allowances for children using actual data on heights and weights in an LDC and DC

	Requirement only	Wasting allowance	Desirable growth allowance
LDC (Asia)	1820	1850	1904
DC (Europe)	2198	2198	2206
Criteria used for children 0–9+ yrs	Actual weight and requirement values	Desirable weight for actual height and requirement values	Desirable weight/age and requirement values

Note: In this table all adolescent and adult weights were kept constant and calculated with a moderate requirement figure for moderate activity. Infection allowances and desirable physical activity allowances are not applied to any of the data, as these allowances are considered later.

5.3 Allowance for childhood infections

The effect of intermittent infections is a major problem for children because as they recover children need a period of catch-up growth simply to make good their weight loss during illness. The cost of this catch-up growth is greater than the energy lost during the illness, so extra provision of energy needs to be made despite the intermittent nature of children's growth with periods of growth faltering and growth spurts.

Table 5.5 shows the increased energy needed for young children to take account of the need for recovery from infection. The 1985 report set out the options for decision making but did not make any recommendations on which course to follow. In the light of the widespread illness and growth faltering in children below the age of 3 years in LDCs, it is suggested that in these countries the energy requirements of children in the first 3 years of life should *routinely* include a factor to allow for catch-up growth.

Table 5.5 The effect of intermittent infections on the energy allowance values in infants and young children up to two years of age in LDCs

Age years	Average weight gain g/kg/day	Predicted usual energy allowances* kcal/kg/day Boys	Girls	Increase in energy needs %	Proposed infection allowance kcal/kg/day Boys	Girls
0.5–0.75	1.83	90	90	14.5		
0.75–1.0	1.15	96	96	8.5	104*	104*
1.0–1.5	0.67	99	103	5.0		
1.5–2.0	0.51	99	103	3.5	103*	109*

* All these values exclude a 5 per cent increase for desirable physical activity which will be dealt with later.

Note: The 1985 report, in specifying general energy requirements, condensed the values for the first year of life into a single figure for boys and girls and took into account the rapidly declining rate of growth from birth to 1 year of age. The estimates generated here presuppose that most babies are not fed additional foods until after 6 months of age and it is only then that they become exposed to any substantial degree to infections with vomiting, diarrhoea and a loss of appetite and the subsequent need for compensatory growth. Therefore, if an average figure for the whole of the first year is needed, then the normal requirement of the first 6 months has to be included with the enhanced need at 104 kcal/kg shown in the table for the second 6 months of infancy. This means that the average energy requirement of infants, i.e. from birth to 1 year, amounts to 103 kcal/kg/day in less developed countries for both sexes. Analyses of growth rates in LDCs suggest that the impact of growth faltering is maximum below the age of 3 years. Therefore the adjustments for infections in the LDCs have only been applied to children in their first 3 years of life.

The impact of this infection allowance on the needs of a single group of girls, aged 18 months, was included in Table 5.2. The effect on a larger group of children aged up to 10 years is shown in Table 5.6. This is exactly comparable to Table 5.3, but now includes the infection allowance. The infection allowance is appreciable, but its effect obviously becomes progressively smaller as the age range to be considered increases. Usually the impact of the infection allowance on the needs of the population below 10 years amounts to an increase of only 6 per cent.

Table 5.6 The impact of infection allowances on the energy needs of a group of children below 10 years of age

	Energy requirement only	Wasting allowance	Desirable growth allowance
LDC no infection allowance	1157	1276	1489
LDC plus infection allowance	—	1352	1577

Note: No data are included for developed countries because these populations are not routinely allocated an infection allowance; their data would be exactly the same as in Table 5.3.

5.4 Allowance for 'desirable' physical activity in children

The 1985 report was unable to provide estimates for the energy requirements of children based on detailed analyses of the energy cost and extent of physical activity. Data for the BMR of children *are* available, and estimates of the cost of tissue growth can be made, but in the absence of information on activity, the estimates of requirements had to be based on the measured *intake* of infants and children. Recent estimates of intake from the developed world suggest that children's intakes have been declining, probably because of the increasingly sedentary lifestyles in affluent communities. This decline in physical activity was considered undesirable by the 1981 Consultation. Measurements of intake of children from LDCs are scant, particularly of those who lead a more active life and either contribute to household work or perform 'paid' child labour as well as, perhaps, walking long distances. Therefore the 1985 report used estimates of intake from developed countries but increased the estimates by 5 per cent to allow for greater physical activity in both developed and less developed countries. This increment is now designated the desirable activity allowance.

   Table 5.7 shows how these desirable physical activity values further amplify the allowance values for young children. The 5 per cent increase has been applied to all children in developed and less developed countries and it is additional to any infection allowance. These figures are suitable for general use in all populations. Table 5.8 specifies the differrence between the requirement and desirable physical activity allowance for older children.

5.5 Energy allowances for adults and adolescents: desirable weight

Men and women may be underweight or overweight for their height and this affects their energy expenditure which, as noted in Section 1.3, can be predicted from their actual weights. Planners, therefore, have to decide whether to use the

Table 5.7 The full allowances for desirable activity and infection suitable for use in young children in less developed countries

Age years	Average weight gain g/kg/day	Predicted usual energy allowance for children in DCs kcal/kg/day		Increase in energy needs to cope with recurrent infection in LDCs %	Proposed infection plus desirable activity allowance for children in LDC kcal/kg/day	
		Boys	Girls		Boys	Girls
0.5–0.75	1.83	95	95	14.5 ⎫	109	109
0.75–1.0	1.15	101	101	8.5 ⎭		
1.0–1.5	0.67	104	108	5.0 ⎫	108	113
1.5–2.0	0.51	104	108	3.5 ⎭		

Note: All these allowances include a desirable physical activity allowance amounting to 5 per cent for children over 6 months of age. From birth to 6 months a similar figure of 5 per cent was added to observed weighed intakes of breast milk to allow for possible experimental errors in test weighing. For simplicity in this manual this 5 per cent figure has been linked to desirable physical activity allowances later in life.

Table 5.8 The effect of the increments in physical activity specified as desirable in the 1985 report

Age (years)	Energy requirement without desirable activity		Energy allowance with desirable activity	
	Males	Females	Males	Females
	kcal/kg/per day			
0+	98	98	103	103
1+	99	103	104	108
2+	99	97	104	102
3+	94	90	99	95
4+	90	87	95	92
5+	87	84	92	88
6+	84	79	88	83
7+	79	72	83	76
8+	73	66	77	69
9+	68	59	72	62
	PAL value			
10+	1.74	1.59	1.76	1.65
11+	1.67	1.55	1.72	1.62
12+	1.61	1.51	1.69	1.60
13+	1.56	1.47	1.67	1.58
14+	1.49	1.46	1.65	1.57
15+	1.44	1.47	1.62	1.54
16+	1.40	1.48	1.60	1.52
17+	1.40	1.50	1.60	1.52
Adults 18–59:				
Activity: Light	1.41	1.42	1.55	1.56
Moderate	1.70	1.56	1.78	1.64
Heavy	2.01	1.73	2.10	1.82
Elderly: 60+	1.40	1.40	1.51	1.56

Note: Energy requirement values were derived from FAO desirable allowances but a lower limit of 1.4 was set. Without this limit male adolescents aged 16+ and 17+ would both have had a requirement value of 1.39. In addition, elderly requirement values would have been 1.32 for males and 1.37 for females. The allowance values exclude the proposed increment for infection.

current weights of their population or those considered 'desirable'. The definition of the most appropriate weight for height is difficult because the only data on which to base this analysis come from developed countries[1] where it has been shown that a body mass index (weight [kg]/height^2 [m]) of about 20.8 in women and 22.0 in men seems to be associated with the best life expectancy within those societies. There is a range in acceptable body mass indices of about ± 10 per cent from the median weight value, the range being 20.1–25.0 in men and 18.7–23.8 in women. In many parts of the world men and women may be underweight or overweight in relation to these standards, but there are little or no data on what constitutes the optimum weight for height under these different circumstances. In the absence of any other information, it is suggested that the 'acceptable' weight range derived from analyses of North American populations should be used if planners wish to apply a normative approach to calculating energy requirements in adults as well as in children.

5.6 Desirable physical activity

This has already been considered in some detail in Chapter 3, where it is made clear that in estimating the energy needs of adults, allowances should be made both for the energy costs of social activities and for short episodes of more intense physical exercise, which will benefit adults engaged in light activity by improving their physical fitness through maintaining the cardiovascular system and musculature. These allowance estimates therefore provide values which may be in excess of those actually observed in many sedentary societies typically found in the developed world.

Detailed analyses of the energy cost of very modest activities in sedentary adults suggest that in both men and women, a PAL value of about 1.4 allows for the cost of being up and about the house in a state of energy balance, but performing little physical work, apart from intermittent moving about for a total of about 3 hours per day. This value of 1.4 times BMR is called the *maintenace requirement* and is the lowest limit set for any usual requirement value in any age group (Table 5.8).

If even this modest activity is limited so that individuals spend almost all day sitting, then the PAL value falls to 1.27. This very low value would permit little or no occupational activity. This issue is considered in more detail in Chapter 3. The impact of the allowances for 'desirable' physical activity vary depending on the age and occupational physical activity patterns of adolescents and adults. These differences are set out in Table 5.8. Clearly the requirement values for those adults involved in light activity are only slightly in excess of the maintenance requirement. This shows how little energy is expended in work, for example in offices, and other light activities in mechanized societies.

5.7 Impact of weight and activity allowances for adults

Table 5.9 produces sets of data based on the observed adult populations and weights from two countries, one an LDC and the other a DC. Here it is apparent that the overall impact of the prescriptive approach to desirable physical activity is to increase energy allowances by 3–13 per cent. In this example adults aged 18–>60 years are being considered as a separate group.

Table 5.9 The effect of desirable weight and activity allowances on the energy requirements of adults in an LDC and a DC

Criteria used	Requirement Actual weight Moderate activity requirement	Desirable activity allowance Actual weight Desirable activity allowance	Desirable weight allowance Fogarty desirable wt for ht moderate activity requirement	Desirable weight & activity allowance Fogarty desirable wt for ht desirable activity allowance
Asian country	2102	2220	2253	2376
European country	2356	2493	2301	2435

Note: In this example the age and sex distribution of the country have been taken as well as information on adult body weights. Appropriate allowances for pregnant women have been included.

Effect of weight allowances at a population level

The impact will depend on the age structure of the population and the degree to which the body weights differ from the standards. The decision to calculate energy allowance by normalizing both growth and adult body weight involve three separate issues:

(a) normalizing weight for age of children (see Section 5.2);

(b) normalizing for adult weight;

(c) projecting the impact of a slow progressive increase in the weight of adults as new generations of taller children enter adulthood.

Obviously in adults one cannot increase height on increasing food intake but in the long term, as the population of children becomes taller, increases in adult height can be expected to occur.

The impact of these allowances is more obvious in LDCs because the problem of weight deficit is usually confined to population groups in LDCs. In several affluent DCs the problem is one of overweight not underweight, so that allowances used as a normalizing process has the opposite effect. In the Asian country chosen in Table 5.10 the energy requirements are increased by 11 per cent to take account of the child and adolescent deficits in weight. Adults, who in this country have body mass indices varying by region from 19 to 21, may be

Table 5.10 The overall effect of specifying allowances for desirable moderate physical activity and the desirable weight for height of children and adults in two nations, one from Asia and one from Europe

	Total population's average kcal/head/day			
	Requirement	Activity allowance	Weight allowance	Allowance for both weight and activity
	Actual weight Actual activity	Actual weight Desirable activity + infection allowance	NCHS/Fogarty desirable wt/ht Actual activity	NCHS/Fogarty desirable wt/ht Desirable activity + infection allowance
Asian country	1820	1927	1941	2053
European country	2198	2328*	2157	2285*

* No infection allowance included since a developed country.

considered to be managing despite being thin: the extra provision of food to cope with the potential impact of having a heavier population of adults of the same height would be over 10 per cent.

5.8 Projecting changes in the height of adults

It is widely recognized that in many countries children are growing faster than 20–30 years ago. They not only reach maturity earlier but have a greater adult stature. This means that young adults are usually taller than their parents; this feature applies to many developed as well as less developed countries as illustrated in Table 5.11 which sets out examples from studies in specific areas of different countries. As noted earlier it is difficult to obtain representative data on heights by age for each country, but it is evident that in some affluent Asian countries there has been a remarkable change in growth rates, suggesting that the difference between the heights of 20- and 60-year-old men and women may amount to as much as 8 cm. In other less developed countries the secular increase in height seems to be small but a change in height is probably occurring in most less developed communities.

For most developed countries the secular trend in height has amounted to 1 cm per decade since 1850, this change being associated with many improvements in environmental conditions and the quality of life[2]. The secular trends may now be slowing in developed countries, with little increase in the stature of young adults over the last 2 decades.

It should be noted that it is normal to find individuals shrinking in height after

Table 5.11 Examples of variations in adult stature and weight by age[3]

Age years	The Philippines Males			Females			Japan Males			Females		
	Wt kg	Ht cm	BMI	Wt kg	Ht cm	BMI	Wt kg	Ht cm	BMI	Wt kg	Ht cm	BMI
20–29+	54.3	162.9	20.4	46.2	151.5	20.1	55.5	161.8	21.2	48.5	150.1	21.5
30–39+	54.9	161.7	21.0	47.1	151.1	20.6	56.4	160.8	21.8	49.0	149.0	22.1
40–49+	53.8	160.8	20.8	47.9	149.6	21.4	56.7	160.0	22.1	50.0	148.2	22.8
50–59+	52.2	160.2	20.3	45.4	149.0	20.4	55.7	159.0	22.0	48.7	147.0	22.5
60–69+	51.8	159.1	20.5	43.9	148.2	20.0	53.7	157.5	21.6	47.0	145.0	22.4
>70	50.3	156.8	20.5	39.4	146.1	18.5	51.2	155.0	21.3	44.0	142.5	21.7

Age	Brazil Northwest Males			Females			Ivory Coast Males			Females		
	Wt	Ht	BMI	Wt	Ht	BMI	Wt	Ht	BMI	Wt	Ht	BMI
20–29+	58.3	165.4	21.3	49.9	154.4	20.9	62.0	167.0	22.2	53.9	157.0	21.9
30–39+	59.6	164.9	21.9	51.6	153.7	21.8	64.9	167.0	23.3	54.8	157.0	22.2
40–49+	59.2	164.7	21.8	52.2	153.3	22.2	65.8	167.0	23.6	55.8	157.0	22.6
50–59+	58.6	164.1	21.8	52.1	152.3	22.5	63.8	166.0	23.2	54.8	156.0	22.5
60–69+	56.7	162.9	21.3	49.8	150.6	22.0	62.8	165.0	23.0	54.8	156.0	22.5
>70	54.9	161.2	20.7	46.6	148.9	21.0	62.8	165.0	23.1	52.8	155.0	22.0

Age	Brazil South Males			Females			France Males			Females		
	Wt	Ht	BMI	Wt	Ht	BMI	Wt	Ht	BMI	Wt	Ht	BMI
20–29+	63.6	170.2	22.0	54.9	158.3	21.9	68.3	173.8	22.6	54.9	160.8	21.2
30–39+	65.8	169.9	22.8	58.1	157.8	23.3	72.0	172.4	24.2	57.4	160.1	22.4
40–49+	66.9	169.6	23.3	60.3	157.6	24.3	73.4	170.7	25.2	60.3	158.7	23.9
50–59+	65.8	169.2	23.0	62.5	156.3	25.6	73.5	169.2	25.7	61.9	157.6	24.9
60–69+	64.1	167.9	22.7	61.7	154.9	25.7	72.3	167.7	25.7	61.1	155.8	25.2
>70	63.0	166.9	22.6	57.3	153.1	24.4	71.0	166.8	25.5	60.2	154.5	25.2

Note: Data relate to surveys of specific regions as provided to FAO and are not necessarily representative of the whole country.

the age of 45 years as their spine shortens, but this amounts to only about 1–3 cm as individuals age from 40 years to 80 years. This will accentuate the differences in height between young adults and the elderly. The cohort effect with the increasing height of young adults by 1 cm per decade means that in developed countries differences between 20- and 70-year-olds should amount to about 5 cm, with an extra 1 cm height difference being attributable to bone and spinal changes with ageing. Thus most of the 6 cm difference is explained by the earlier secular differential growth rates of children.

5.9 Effects of height and weight changes during ageing and as a reflection of the cohort effects

Allowance should ideally be made in each country's analysis for the differences in height of adults of different ages, but this is difficult to do on a systematic basis. The provisional figures in Appendix 2 pre-suppose, in the absence of height data by age and sex, that women of all ages have a height equivalent to that of 18-year-old girls and that men's height is 7 per cent greater at all ages. No allowance has therefore been made for the secular increase in height in these tables. Were this allowance to be made in developed countries, the effect would be to reduce energy requirements by 3.3 per cent if it is assumed that the body mass index of adults of all ages is the same. In practice, adults in developed countries show an increase in body mass index (BMI) until they are in their 5th or 6th decade of life, when the BMI stabilizes before declining slightly in the 8th decade. If one is dealing with actual weight and height data then the decrease in height with age is more than complemented for by the increase in weight and body mass index with age.

Yet any increase in weight with its effect on increasing basal energy needs is also counteracted by the observed fall in basal energy needs per kg body weight with age. This change reflects the disproportionate loss of lean tissue and the excessive deposition of body fat in older adults; basal metabolic needs depend on the mass of lean tissue, so the BMR expressed as kcal per day tends to fall with age despite the increase in body weight. This is illustrated in Table 5.12.

A further factor affecting total population energy requirements in the developed world is the well-recognized fall in both the intensity and duration of physical activity in middle age. The progressive fall in physical activity extends

Table 5.12 Impact of the cohort effect on height, of increasing weight for height with age, and the effect of reduced spontaneous leisure time activity in adults from an LDC and a DC on the average national energy requirements

Assumptions	Community in LDC e.g. Ivory Coast kcal/head/day	Community in DC e.g. Japan kcal/head/day
1. Height: no cohort effect (a) BMI constant from age 18 years (b) Activity constant, e.g. moderate desirable activity for whole adult population	2090	2212
2. Actual differences in height of adults (a) BMI constant from age 18 years (b) Activity constant, e.g. moderate desirable activity for the whole adult population	2087	2183
3. Actual changes in heights and weights of adults. Activity constant, e.g. moderate desirable activity for whole adult population	2077	2178
4. Actual changes in height, weight and estimated decline in activity with age: Moderate desirable activity for 100% 18–29 yrs 50% light 50% moderate desirable } for 30–59 yrs Light desirable activity for 100% elderly	2036	2104

Note: For this table the estimated population of each country was held constant in all the calculations with there being no change in the estimated requirements of children in any of the estimates. The effects relate exclusively, therefore, to the different assumptions made on adult weights, heights, and activities.

into old age. The combination of the falling activity and BMR decline with age is also apparent in the limited data available from developing countries[4] but detailed studies of physical activities in older people in less developed countries are very limited.

The relationship between age and energy requirements expressed on a cross-sectional basis is therefore complicated and Table 5.12 presents a synopsis of the likely impact of differences in height, weight, and leisure time activities in adults in several decades of life. This table shows that in the Ivory Coast, where there is little difference between the heights of young and old adults, the effect on energy requirements is negligible, whereas in Japan a difference in height theoretically reduces the requirement by 29 kcal/day because of the lower energy requirements of the shorter older Japanese. However, this effect is compensated for by the enhanced weight of the middle-aged Japanese, with an increase in average national requirements occurring despite the height differences. The differences in adult weight and height, however, may be less important than the change in physical activity which tends to dominate the picture.

5.10 Allowances in pregnancy

When the FAO/WHO/UNU consultation met in 1981 there were only a limited number of studies which allowed an assessment of energy requirements in pregnancy; reliance was placed on estimates of the energy laid down in the mother and foetus, and the cost of synthesizing and maintaining the new tissues during pregnancy. On this basis it was concluded that 80 000 kcal was a reasonable total figure for the extra demand for the whole of pregnancy.

If women's metabolic and physical activity remained constant throughout pregnancy then the additional 80 000 kcal represents an increment in energy requirements. Although in 1981 there was some evidence that women's BMR might not increase as much as expected and that physical activity fell, a cautious policy was adopted so that the 80 000 kcal value was taken as an extra demand. This amounts to an energy allowance of 285 kcal daily throughout pregnancy. Alternatively, the allowance was reduced to 200 kcal per day if healthy women reduce their activity.

Since 1981 there has been a surge in studies on energy metabolism in pregnancy, almost all of which show that the expected increase in intake of 200 or 285 kcal does not occur. Indeed, almost all studies show no increase at all in food intake. Measurements of BMR show modest increases in late pregnancy, with a reduction in the metabolic response to food and a tendency to reduce physical activity in late pregnancy. On this basis it has been suggested that a modest increase in intake of 100 kcal per day might have been barely detectable but would have covered the observed increase in basal metabolic rate. On this basis we can consider three values:

Pregnancy requirement: 100 kcal/day

Pregnancy sedentary allowance: 200 kcal/day
Pregnancy active allowance: 285 kcal/day.

5.11 Assessing the pregnancy rate in a population

The simplest way of dealing with this is to take the total number of annual births (x) recorded for a country and assume that $0.75x$ women are pregnant at any one time throughout the year. This therefore makes an allowance for pregnancy lasting 9 rather than 12 months. The value of 0.75 for converting birth rates to pregnancy rates neglects the number of miscarriages, but this effect is counteracted by also neglecting the number of twins produced. More complicated methods are available for calculating the pregnancy rate for each year when the annual birth rate is changing rapidly, but this requires expert statistical help. Having established the number of women who are pregnant at any one time in a year, one can then apply the requirement or either allowance figure as though it applies for each day of the year.

The impact of the pregnancy allowance values is shown in Table 5.13, where it is evident that on a population basis the effect is small. If, however, one were dealing with a group of young women then the total estimated food allowances would be appreciably different.

Table 5.13 Impact of energy allowances for pregnancy on the average population allowance values

	Number of births	Number pregnant	Requirement of non-pregnant population	Pregnancy requirement	Pregnancy desirable sedentary allowance	Pregnancy desirable activity allowance
LDC (Asia)	22 962 000	17 221 500	1919	1921 (0.10)	1923 (0.21)	1925 (0.31)
DC (Europe)	640 000	480 000	2380	2381 (0.04)	2381 (0.04)	2382 (0.08)

( ) Allowance for pregnant population as a percentage of total population allowances.

5.12 Allowances for lactation

Conventionally, energy requirement calculations undertaken by FAO have dealt with this topic very simply by specifying that a mother and her infant would be considered as a unit. The food needs of the breast-fed child are then taken as that which the mother would have to contribute by eating more. Thus by specifying the energy needs of the breast-fed child and including them in estimates of population energy requirements, one does not add on any extra for the lactating mother.

In practice, the 1985 report specified the cost of milk synthesis at 20 per cent of the milk energy secreted, estimated median milk production rates at each phase of lactation, and recognized that the average woman would start lactation with an extra store of 36 000 kcal as fat reserves. This reserve provides 200 kcal per day for the first 6 months of lactation, so the additional energy requirement for the energy cost of lactation was reduced from 700 kcal/day on average to 500 kcal/day. Later in lactation the energy requirement would have to be met by extra food being eaten. Some studies suggest that lactating women eat enough additional food to compensate for the estimated cost of lactation, whereas other studies show little or no increased energy intake. Studies on the metabolic efficiency of the lactating woman and her physical activity pattern are sparse, so it is not clear what her true average requirement would be. Furthermore, the length of breast feeding, the time and extent of supplementation for the infant and the energy demands of child care are all areas of uncertainty in different countries. This is the justification for FAO's choice of a simplified approach whereby an infant's energy needs are specified, and this is included in a household or population's food needs while neglecting the energy cost of milk synthesis or metabolic changes in the mother. Behavioural alterations in physical activity patterns can be monitored in lactating women, who would then be included in the general analysis of women's activity patterns.

References

[1]Bray, G. A., Ed. (1979). *Obesity in America*. Proceedings of the 2nd Fogarty International Center Conference on Obesity, Report No. 79, Washington, DC, Department of Health, Education and Welfare. Based on: 'Mortality among overweight men and women', Statistical Bulletin 41, New York, Metropolitan Life Insurance Co., 1960.

[2]Tanner, J. M. (1962). *Growth at Adolescence*, 2nd Edn. Blackwell Scientific Publications, Oxford.

[3]Food Policy and Nutrition Division, Economic and Social Policy Dept. (1984). *Method for estimating energy requirements by countries. Comparison with previous estimations.* Informal gathering on methodology of the Fifth World Food Survey, 12–14 March, FAO, Rome.

[4]Shetty, P. S., Soares, M. J., and Sheela, M. L. (1986). *Basal metabolic rate of South Indians.* Report to FAO from the Nutrition Research Centre, St. John's Medical College, Bangalore, S. India.

6 Effect of different assumptions on estimated allowances

In previous chapters an attempt has been made to indicate how a nutritionist or planner can choose different assumptions regarding the population characteristics for calculating the energy needs of the population. It is difficult, however, to know how important each of these factors is in determining the final estimate. The purpose of this chapter is to illustrate the likely extent to which each of the many factors contribute to the average allowance figure. It should be recognized that many of these factors interact and either add or subtract from the final value. In practice there are also correlations between factors, so it is difficult to disentangle the impact of a single feature.

6.1 Importance of a population's weight: actual and 'desirable'

Body weight is a major determinant of energy requirements, as illustrated in Fig. 6.1. In this figure the actual weights of children and adults from a single country in four continents have been chosen but the activity levels of the four countries have been held constant simply to illustrate the weight effect. It is evident that there are appreciable differences in the energy allowances and that this is apparent particularly for children in Asian countries. In one country requirements were increased by 8 per cent and 27 per cent for children aged 0–17 + years, depending on whether allowances were made for either deficits in weight for height or in weight for age, i.e. reflecting the extent of wasting and stunting. Stunting is therefore an important contributor to the reduced energy needs of some countries and a normal growth allowance increases food estimates considerably.

If the overall effect of body weight at all ages is estimated from this figure, then, with adult Asian females weighing 43 kg, the national average allowance is 1925 kcal per head; in an African country with adult females of 52 kg national allowances are 2063 kcal per head; in a European country the values are 54 kg and 2328 kcal per head; in North America 57 kg and 2387 kcal per head. This effect of weight has been given inadequate emphasis in many attempts to develop energy allowance figures.

Figure 6.1 illustrates the importance of specifying whether the actual weights are to be used in estimating allowances or whether the planner wishes to take

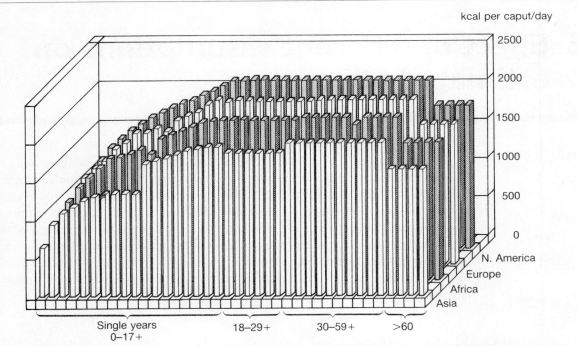

kcal per caput/day

Fig. 6.1 Effect of weight differences on the average energy allowances of females living in one of four continents.

account of prospective changes in weight. This adjustment must be specified within a certain time frame because, as noted in Chapter 5, societies which are becoming more affluent are showing surprisingly rapid changes in children's growth. Adult stature is also increasing steadily. The children's growth curves from affluent subgroups within many racial groups are surprisingly similar, and are close to the NCHS standards set out in simplified form in Appendix 3. This finding implies that, as living conditions improve, there will be a tendency for most populations in the less developed world to attain growth curves similar to those found in Europe and North America. This in turn means that on a long-term basis, the same size of population will need more food simply to sustain its greater stature as the benefits of improved hygiene, social conditions, and nutrition lead to greater growth rates in the children. This is of great importance for long-term projections. In some countries, e.g. Japan, there has been a very rapid change in heights. Similar effects are seen on a more modest scale in many parts of Europe where height differences are still evident in different social classes, although the heights of the most affluent in society have not changed appreciably for over a century.

Note that Fig. 6.1 refers to actual body weights observed in children and women from a country for each of Asia, Africa, Europe, and North America. For the comparison, activity patterns have been arbitrarily set at moderate levels with the inclusion of desirable activities. The population's total need for energy is not given since this will, in part, reflect differences in the different sizes of the population and its age structure. This graph simply depicts the age- and weight-related requirements. For children each column illustrates the individual requirement (kcal per day) for single years. The per caput values for adult women are shown using six, ten and four columns, respectively, for the age groups 18–29+, 30–59+ and over 60 years. The per caput values are constant for each adult range and do not show the expected gradual decline in requirements on an age basis, since this is not given in the standard equations for each age band.

6.2 Desirable versus actual weight of children and adults

Short-term adjustments to allow desirable weights for heights of children and adults to be achieved have been dealt with in Chapter 5, but the impact of this effect on the average energy requirement is illustrated in Fig. 6.2. Clearly an adjustment for deficit in weight for height is important in the Asian community (1925–2050 kcal/caput/day), whereas there is a very small decrease in the other three countries in keeping with the excess weights in these communities. African and Asian countries would have an appreciable further increase in requirements were the children's weights to be adjusted to the NCHS desirable weights for age. If one expects in the long-term the women of the Asian community to match the heights of 18-year-old girls on the NCHS standard scale, with men having a height 7 per cent in excess of this figure, then clearly there is a substantial long-

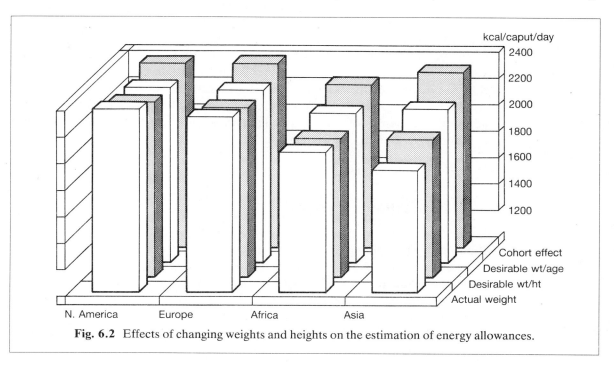

Fig. 6.2 Effects of changing weights and heights on the estimation of energy allowances.

term difference to be seen in the average energy requirement figures. These will increase to about 2327 kcal/caput/day on average. Therefore it is important to decide not only the size of the population projection (see below), but also whether heights of children and adult populations are going to increase at an appreciable rate.

Note that in Fig. 6.2 activities have been maintained constant at moderate levels. The same countries given in Fig. 6.1 have been chosen using, for the front row of histograms, the actual body weights of children and adults. The further histograms illustrate the differences to be expected with each new body weight. The columns have been truncated at their bases (by 1000 kcal) in order to highlight the differences in per caput requirements. The kcal/caput/day values for the actual weights are 2387, 2328, 2063, and 1925 for the North American, European, African, and Asian countries, respectively; corrections to desirable weights for height give 2328, 2285, 2053, and 2050 kcal/caput/day; adding an increment to provide children and adolescents with a requirement which allows for a desirable weight for age raises the values to 2321, 2303, 2133, and 2163 kcal/caput/day. The final row of values are 2387, 2392, 2233 and 2327.

6.3 Population structure

Figure 6.3 shows that in Type 1 populations with a predominantly old age structure the choice of activity values in the adult has a substantial impact on the overall national need. In a Type 4 population the effect is diminished, since children make a greater contribution to the overall national energy demands and they are assumed not to change their activity in parallel with adults. In practice some increase is likely; this will minimize the interaction between population structure and physical activity on overall energy requirements. An hypothetical total population of 100 million was used, with actual body weights and total birth rates taken from data for a West African country. The age structures were then adjusted to comply with the four populations shown in Fig. 4.3. The four age pyramid types are discussed in Chapter 4.

6.4 Population increases

When projecting potential changes in national energy needs, a crucial step is the estimation of the likely population changes. Figure 6.4 shows the impact of population changes in a West African country projected from 1950 at 10-year intervals until 2020. The actual weights of the population are held constant throughout the sequential analysis, as are the physical activity factors. Therefore the effects only relate to the changing population numbers. There is a profound change in the total energy requirement of the population in proportion to the

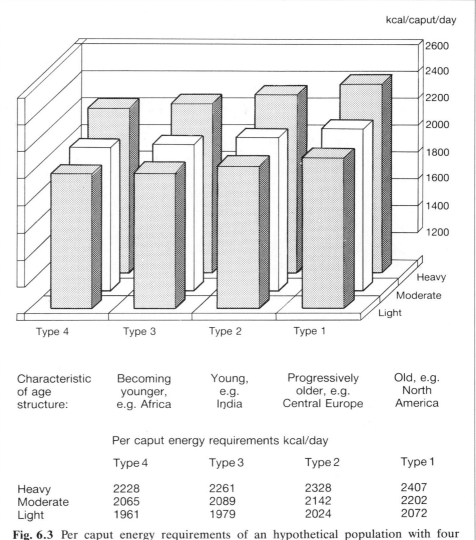

Characteristic of age structure:	Becoming younger, e.g. Africa	Young, e.g. India	Progressively older, e.g. Central Europe	Old, e.g. North America

Per caput energy requirements kcal/day

	Type 4	Type 3	Type 2	Type 1
Heavy	2228	2261	2328	2407
Moderate	2065	2089	2142	2202
Light	1961	1979	2024	2072

Fig. 6.3 Per caput energy requirements of an hypothetical population with four different age and sex structures.

changing numbers in each age group. *These effects of the population size are so great that they are likely to overwhelm all other considerations relating to total country level energy requirements.*

The total rather than the per caput requirement for boys and men is illustrated. The different patterns therefore reflect the estimated changes in the population age structure. For purposes of clarity, values for adults have been converted to *approximate* yearly population requirements by dividing the summed energy needs for the adult age ranges by the number of years covered by that age band. For the over-60-year age group three columns are given, with a progressive but proportionate decline to reflect the likely falls in total energy needs of the age

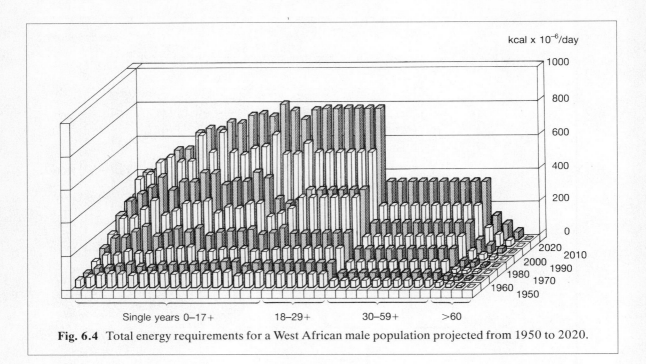

kcal x 10^{-6}/day

Single years 0–17+   18–29+   30–59+   >60

Fig. 6.4 Total energy requirements for a West African male population projected from 1950 to 2020.

group up to an arbitrary limit of 75 years. If stature increases and body weight becomes greater, then this will accentuate the increase in total demand for food from 1950 to 2020.

6.5 Urbanization

Figure 6.5 illustrates the effects which one might anticipate from changes in urbanization in less developed and developed countries. In this case the changes are relatively small and because of current assumptions, they only affect the adult values. The impact of changes in adult energy allowances will be modest in communities with a large proportion of children. This may, however, be simply a theoretical evaluation because, in countries with a high proportion of children, many of them will be working and thus alter the true energy demands of the population.

6.6 Interactions

The effects shown above can obviously interact and this has to be taken into account, particularly when making projections about future energy needs. Some

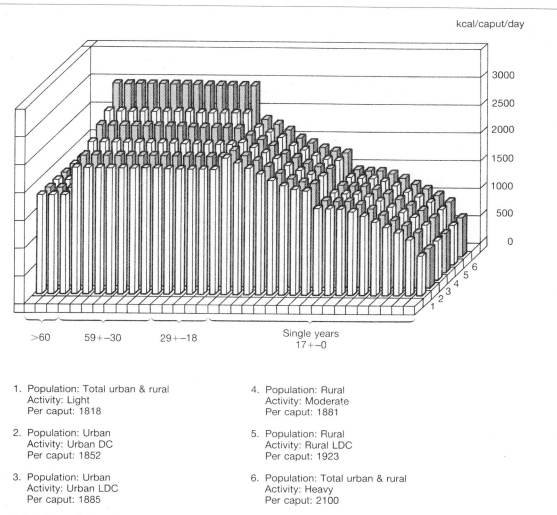

kcal/caput/day

>60   59+–30   29+–18   Single years 17+–0

1. Population: Total urban & rural
   Activity: Light
   Per caput: 1818

2. Population: Urban
   Activity: Urban DC
   Per caput: 1852

3. Population: Urban
   Activity: Urban LDC
   Per caput: 1885

4. Population: Rural
   Activity: Moderate
   Per caput: 1881

5. Population: Rural
   Activity: Rural LDC
   Per caput: 1923

6. Population: Total urban & rural
   Activity: Heavy
   Per caput: 2100

Fig. 6.5 Effect of changing assumptions about activities for working adult males in an Asian country. The same population size was used in conjunction with the same body weights of children and adults. All calculations included an allowance for young children. The assumptions used in the graph, with per caput and requirement values, are given below.

of the factors are correlated since urbanization, for example, is a feature of affluent developed countries, where the population tends to be both tall and with little or no deficit in weight for height. Berio et al.[1] have analysed these interactions and observed that the most important factors affecting average national energy allowances were the anthropometry and the age structure of the population. The anthropometry accounted for 49 per cent of the observed and explained variance in national energy needs. The age structure accounted for 35 per cent of the variance. Compared with these two factors, urbanization seemed to account for only 0.5 per cent of the observed variance.

These findings do not, of course, deny the importance of urbanization in reducing energy needs. They simply signify that, in general, those communities which have become urbanized have either changed their population structure, or had appreciable weight increases in the population to compensate for a fall in physical activity in urban dwellers. These two effects of weight and age then dominate the picture. In individual countries, however, this general picture may not hold.

6.7 Conclusions

We may conclude that in population projections one needs to take account of a number of interacting factors. The dominant effects are:

1. The weight of the children and adults.
2. Alterations in physical activity.
3. Changes in the population structure.
4. Urbanization effects which may be complicated, with little or no change in energy needs in many countries.
5. The population size. This is of overwhelming importance when considering total population food needs.

Reference

[1]Bério, A. J., François, P. and Perissé, J. (1985). *Population dynamics, anthropometry and urbanisation: their relationship to human environment.* Paper for the International Union of Scientific Study of Population (IUSSP), 5–12 June, Florence, Italy. Published in Food and Nutrition, **11**, (1), 1985 (FAO).

7 Adaptation and survival on low intakes

In the 5th World Food Survey an analysis was made of the incidence of under-nutrition by comparing data obtained on food intake with the estimates of energy expenditure. It is evident that people can remain in energy balance on low energy intakes by simply doing very little. By lying in bed all day it can be shown, for example, that a PAL value of about 1.2 is the cost of maintaining weight in the short term—thereafter progressive muscular wasting and bone loss not only reflect deteriorating health, but may also lead to weight loss. The question then is one of defining 'malnutrition' in some meaningful way. To measure malnutrition in a population there is a need to specify two features: (1) Abnormally low body weights, low growth rates or low birth weights. These are taken to depend on a deficit in food intake and, for the purposes of the 5th World Food Survey[1], on a deficit in energy intake; (2) Abnormally low rates of physical activity.

The 5th World Food Survey categorizes the ability of people to adapt to a low energy intake into three components:

1. *Social/behavioural adaptation*, i.e. where individuals may modify their activities in response to a fall in food intake. This could occur without any weight change.

2. *Biological adaptation*, i.e. associated with a change in body weight and energy stores. This occurs when a reduction in activity fails to compensate completely for the decline in energy intake. Therefore body energy stores will be reduced with a fall in body weight. As the loss of energy and body weight progresses, so the amount of metabolically active lean tissue declines. This in turn reduces the BMR, usually the principle component of energy expenditure, so the individual progressively compensates for the energy deficit as the fall in BMR buffers the deficit. Eventually the body comes back into energy balance at a lower body weight and with a lower BMR. The body adapts to increases in intake in a similar manner. This mechanism of adaptation is well recognized and now widely accepted.

3. *Metabolic adaptation*: should the body be able to modify its metabolic efficiency without change in weight or activity, then this would be a further mechanism of compensation and not constitute a hazard to the individual.

The FAO/WHO/UNU Expert Committee in 1981 considered that there was some evidence for the social/behavioural adaptation, accepted that biological adaptation does occur, but rejected metabolic adaptation as a process to be taken into account in its policy-making. Therefore the principle issues in considering the energy requirements of a 'healthy population' are the definition of the levels of adult body weight and children's growth which are acceptable, and the levels of physical activity which are appropriate to the well-being both of the individuals and of society. Clearly, deciding on these levels is not easy and it is in this area of policy where some experts will have different views from those proposed by the 1981 Expert Consultation. This manual provides the planner with the option of making independent judgements on these issues.

7.1 Acceptable body weights

The distinction between the terms 'acceptable' and 'desirable' is important. The word 'acceptable' reflects a cautious view of what weight may be appropriate, whereas the word 'desirable' clearly specifies the need to achieve that weight and thereby avoid some risk. The 1985 report used the word 'desirable', but when dealing with many different countries where information on the optimum weight is not available, it may be better to refer to those North American values as 'acceptable'.

Choosing acceptable weights is not easy because the criteria for specifying these weights are unclear. The 1985 report used the only readily usable weight range as proposed by the Fogarty Committee in the USA[2]. The tables of acceptable weight were originally chosen from life expectancy studies of adults taking out insurance policies in the United States before the Second World War. A useful simplification is the expression of weight for height as the Body Mass Index (BMI). This index was chosen because it has approximately the same value for short, medium height, and tall groups. The index therefore simplifies the approach to calculating accurate weights for groups of different height. The average BMI is taken to be 22.0 in men and 20.8 in women. Table 7.1 sets out the BMIs chosen by the 1985 FAO/WHO/UNU report.

Table 7.1 The Body Mass Indices for adults based on American insurance statistics

	Men	Women
Acceptable range of body mass index	20.1–25.0	18.7–23.8

Note: Body Mass Index is calculated as $\text{BMI} = \dfrac{\text{Weight in kg}}{\text{Height}^2 \text{ in m}}$

Clearly, different societies may experience a long life expectancy at different weights from those based on North American statistics, but the difficulty is that there are no data from other countries which provide any basis for adjusting these figures. In many parts of Asia the BMI of adults is 18 or 19, but whether these lower BMIs are an advantage or a disadvantage in these societies is unknown. If physiological tests of work capacity are performed where the total energy output of the body is measured, then on average groups with a BMI of 24 will have a greater capacity for physical work than those with a BMI of 19–20. These physiological tests may not, however, be a realistic test of the ability of men and women to work under practical conditions and the advantage of an enhanced work capacity must be balanced by the recognition that the non-occupational energy requirements of adults with a BMI of 18 is substantially less than those with a BMI of 25. This is illustrated in Table 7.2, which shows what a profound effect both body weight and body height have on the estimated energy requirements of the population.

Table 7.2 shows a wide range in acceptable body weight and the assumption within the 1985 report is that any weight within this range is appropriate. Thus, for men who are 165 cm tall, the difference in maintenance energy requirements of 2117 kcal at a BMI of 20 compared with a BMI of 25 amounts to 292 kcal a day or a reduction of 12 per cent in energy requirements should the same population be at a BMI of 20 rather than 25. This is therefore a biological adaptation of substantial importance and accepted as costless adaptation by the 1985 report. If planners consider that they should simply plan on the basis of the observed BMI of, for example, 18 in their society, then this would signify a fall in energy requirements of about 15 per cent in the adult population from that expected at a BMI of 25, or about 10–12 per cent from a midpoint BMI value of about 22.

Table 7.2 Maintenance Requirements (MR) in kcal per day of adults of different weights

Body Mass Index:	Men						Women					
	18		22		25		18		22		25	
Height (m)	Wt	MR	Wt	MR	Wt	MR	Wt	MR	Wt	MR	Wt	MR
1.50	40.5	1818	49.5	2011	56.3	2157	40.5	1528	49.5	1713	56.3	1853
1.65	49.0	2000	59.9	2237	68.1	2409	49.0	1703	59.9	1927	68.1	2096
1.80	58.3	2199	71.3	2478	81.0	2686	58.3	1894	71.3	2162	81.0	2361

Maintenance requirements are defined for the purposes of this calculation as 1.4 BMR and the adults are taken to be aged 18–29+ years of age. Body weight is expressed as kg.

The effect of height is also apparent in Table 7.2. Clearly, the energy requirements of short groups are very much less than tall groups when both are of the same body proportions. Thus very short adults living for example, in parts of Asia, will have a maintenance energy requirement which is nearly 20 per cent less than that of tall people of equivalent activity living, for example, in parts of Africa. On this basis it is recognized that short adult stature could be considered as a form of long-term intergenerational biological adaptation. Thus, in theory, short people have a biological advantage when agricultural land is relatively unproductive. This concept has been seen as of major significance by some nutritionists and planners, who emphasize that one should not necessarily regard the progressive increase in height and size of a population as a welcome development.

Yet, intrinsic to nutritional thinking is that children should grow 'well', i.e. at an optimum rate. The 'optimum' rate is usually considered to be the highest possible rate, with the implication that slower rates of growth are in some way disadvantageous because the children are not growing to their full genetic potential. The relationship between early growth rates and long-term health is poorly understood and it is possible, and indeed likely, that *a range* of growth rates of populations will eventually be seen as without cost to the welfare of the individuals or populations. For the present, the 1985 report specifies the NCHS reference tables as likely to reflect the growth patterns of well fed children free of infection, whatever their racial background. Studies do suggest that some Asian groups and unusual African populations of pygmies do not attain these standards even when, as in the Asian case, they are well fed. Nevertheless, detailed analyses of this issue are not available and it would be unwise to assume that populations are unable to achieve the NCHS scales of growth. If, therefore, the environmental needs of children are satisfied, we can expect in the future to see a progressive increase in adult stature and therefore a progressive increase in the energy needs of the population. The increase in adult stature leads to age-related changes in average adult height as illustrated in Chapter 5. Chapter 6 examined the likely interactions of the principle determinants of energy requirements on the final average requirement value.

7.2 Metabolic adaptation

This is still a contentious issue, not because metabolic adaptation does not occur, but simply because it is unclear whether it can occur in healthy individuals of an appropriate weight who are not also changing their body weight. If the energetic efficiency of tissues can be altered without detriment to the functioning of the individual and without being associated with progressive weight changes or an unsatisfactory degree of underweight or overweight, then this would affect one's analysis of a population's energy requirements. Unfortunately, at present there is little information on this point because all studies so far have observed appreciable weight changes in association with an alteration in the body's metabolic efficiency. Furthermore, the intraindividual changes in energy metabolism on a standard diet, or on a daily fluctuating diet, or during sustained over or underfeeding have all induced only small changes in daily total energy expenditure equivalent to 10–20 per cent of the amount of food energy removed or added. A 100 per cent change in expenditure would be needed to compensate fully for the altered intake.

The 1985 report, which was based on the limited data available up to 1981, was unable to accept metabolic adaptation as a feature of costless adaptation and therefore did not take it into account in their calculations. The 5th World Food Survey noted, however, that Sukatme, on the basis of statistical analyses, proposed that *intra*individual changes in energy expenditure might be substantial and indeed account for the *inter*individual differences in intake and/or expenditure. Whilst there is no physiological evidence for this intraindividual variation in metabolic efficiency under normal, everyday conditions, there is no conclusive proof that it does not occur. Therefore, when estimating undernutrition, the 5th World Food Survey took as one option a cut-off point for normality which corresponded to the low end of the range of interindividual differences in energy requirements, i.e. -20 per cent of the mean. Since the maintenence energy requirement used was 1.4 BMR, the cut-off point was therefore taken to be 1.2 BMR. Only when intakes were below this level would malnutrition be specified as occurring. If one rejects the concept of costless adaptation, then clearly, by specifying that an intake equivalent to just over 1.2 BMR does not signify malnutrition, one is ignoring the fact that at this intake the individuals must adapt by either reducing their physical activity or their body weight below the desirable levels. For these reasons the 5th World Food Survey chose to define two cut-off points corresponding to 1.2 BMR and 1.4 BMR. The first reflects the most conservative estimate of the likely incidence of malnutrition, with its assumption of costless metabolic adaptation, whereas the 1.4 value was that favoured by the 1981 Expert Consultation.

7.3 'Maintenance' and 'survival' requirements

The 1985 report does not specify maintenance requirement figures but the 5th World Food Survey, on the basis of experimental work published since then, proposes that the maintenance requirement should be taken as 1.4 BMR. This incorporates about 3 hours of activity while standing and includes washing and dressing activities, but not occupational or socially desirable activity.

If an hour's activity is removed, then calorimetric calculations show that this level of inactivity amounts to 1.27 BMR. This simply allows the individual enough energy to get out of bed, dress and eat, but presupposes that others grow or purchase food, prepare it and clear up after the meals. This value may be considered as the '*inactive*' requirement to signify total dependence such as might apply to immobile people living in retirement homes and taking no exercise. This requirement still allows enough energy for people to get out of bed during the day, to read, write, talk and engage in other sedentary activities. It does not allow for the activity required to maintain physical fitness or muscular tone and strength. It cannot therefore be considered a 'desirable' level of physical activity.

If people are confined to bed then their energy requirements may fall further and amount to about 1.2 BMR, which may be termed the *survival* requirement. The provision of food that would supply this amount of energy could only be seen as an emergency measure, as in the acute phase of famine relief, and the body will deteriorate progressively at this intake.

Figure 7.1 summarizes the different requirement figures for men and women and indicates how important small increments in the PAL values are for making energy available for different activities. This also means that when assessing food

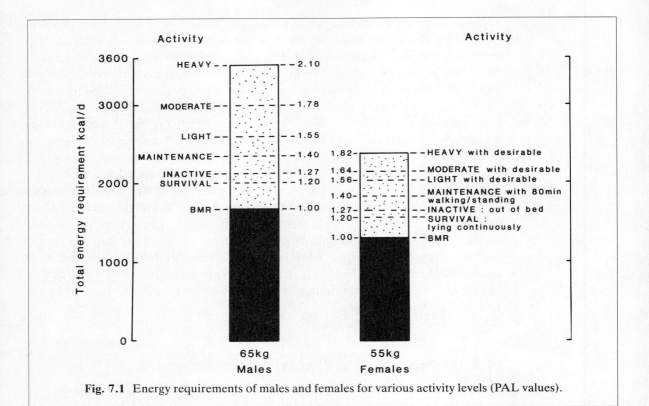

Fig. 7.1 Energy requirements of males and females for various activity levels (PAL values).

supplies or energy requirements, it has to be recognized that small differences in the calculated PAL values make a substantial difference to what the population can do physically. It is also very important to have accurate information on the body size of the adults as well as children because this will make a very great difference to the estimated average requirement expressed in kcal/head/day.

The values chosen in Fig. 7.1 are average values for groups of men and women sustaining an energy expenditure at the specified level. In practice, individuals vary in their energy requirement at these levels with a coefficient of variation of about ± 10 per cent when the requirement is expressed in kcal per day at a specified weight, age and activity. Thus, although there is a twofold range in energy expenditure (kcal/day) in adults of different sizes, the range reduces substantially when the level of physical activity and body weight is specified. This effect reflects the fact that body weight, rather than interindividual differences in metabolic rate or metabolic efficiency, dominates the requirement figure.

References

[1]Food and Agriculture Organisation of the United Nations (1987). *Fifth World Food Survey,* FAO, Rome.

[2]Bray, G. A. (ed.) (1979). *Obesity in America.* Proceedings of the 2nd Fogarty International Centre Conference on Obesity, Report No. 79, Department of Health, Education and Welfare, Washington, DC.

8 Special applications

8.1 Emergency feeding

Two $5\frac{1}{4}''$ disks are included with this manual which allow the calculation of energy requirements and allowances on a personal computer. The User Guide is given in Appendix 6. Estimates of energy allowances can be made within a matter of minutes once some essential information is obtained or assumed. Where there is insufficient time for this approach, a cruder procedure can be used. Thus it should be possible to do a headcount of refugee numbers, assess the relative proportion of children and adults, and then measure the body weights of a small sample of both children and adults. A simple table is then made to provide estimates of the group's food needs. Clearly, in emergency feeding conditions, it is essential to provide sufficient food for maintaining life, which therefore means supplying a minimum of 1.4 times the estimated BMR of the group.

Table 8.1 provides an example of emergency daily energy requirements calculated by the present method. In an African population with a dominance of children who, on average, have not grown above the 20th centile of weight for age, the temporary maintenance requirement is about 1600 kcal per head. This arbitrarily assumes a BMI of 19 for adults, and body weights of 52 kg for men and 44 kg for women. As in other circumstances, climate will not normally influence

Table 8.1 An example of the calculation of energy requirements in an emergency (kcal/cap/day)

Group	Temporary maintenance requirement*	Allowance for the long-term*
0–1	750	790
2–3	1090	1150
4–6	1300	1370
7–9	1430	1500
10+ (male)	1950	2360
10+ (female)	1630	1980
Pregnant women	1730	2180
Average per caput per day	1590	1880

Selected criteria for this example:

Infants and children	Requirement	Desirable activity level
Adults and adolescents	1.4 BMR	1.7 BMR
Pregnant women	Requirement allowance	Sedentary allowance

*All values rounded.

energy needs. This does not mean that a cold environment will not lead to deaths of exposed children and adults; in such circumstances extra feeding of these populations may affect the death rate, if it permits greater physical activity to keep warm.

Clearly the daily per caput requirements will vary according to the weights and structure of the population. This method allows the estimation of energy requirements according to specific local needs in contrast to widely used fixed allowances. The issues of environmental temperature and levels of allowance were discussed in detail at a recent conference on nutrition in times of disaster.[1]

Example of calculating the tonnage of food needed

Rations can be estimated as grain equivalents where 2000 kcal can be provided by 450 g grain per day plus 50 g oil per day, i.e. 15 kg food per month.

Thus 5000 people need: 67.5 MT grain per month
and 7.5 MT oil per month
= 75 MT of food per month

Details of emergency feeding changes and how to handle emergencies are given in a practical guide to Drought Relief in Ethiopia.[2]

8.2 Assessing the energy allowances of special groups

It will be clear that calculating the energy allowances of special groups will usually be a much simpler procedure than assessing national energy needs because a group often has a more uniform activity pattern and may well be from only a single age and sex group. Examples of the estimation of energy requirements of adolescents and the elderly will be used to illustrate the way in which this can be handled.

Table 8.2 provides the data for the collation of the activities of young adolescent boys at school and during their free time. As noted by the 1985 report, evidence on the activities of adolescent boys and girls is limited. Table 8.3 summarizes Finnish experience on activity patterns, the energy cost of these activities being derived from new data.[3] These activity patterns are the observed patterns and not the prescriptive activity patterns suggested by the 1985 report. The PAL values come much closer to those implied by the measured energy intake data and by the proposed PAL values minus the desirable component suggested by the 1985 report (see Section 1.1.2).

A summary of some Italian studies is given in Tables 8.5 and 3.1 to show how simple the collation of activities can be in the elderly. It is evident that men and women sustain their activity into late old age provided they are well enough to live in their own homes, women tending to retain their mobility into the 8th decade. After 80 years there does seem to be some decline in activity and the amount of moving seems to be greater if people live at home rather than in retirement homes. It is not of course possible to assess whether the less mobile were selected for occupying a retirement home or whether the environment conditioned their behaviour. Nevertheless, Table 3.1 does emphasize the need to take into account the social circumstances of special groups when attempting to estimate their energy requirements. When the PAL values were calculated for 421 elderly men and 451 elderly women in Italy, they proved to be 1.59 for the

Table 8.2 The activity patterns of boys aged 10 and 11 years (A) at school and (B) at home in Finland in 1979[4]

A. School Day	Duration % of day	Time min	Energy Index	Energy expenditure kcal
Column	1	2	3	
Activity group				
School duties	24.6	354	2.38	752
Travel to school	2.4	36	3.44	111
Assistance labour force	0.2	3	2.33	6
Housework	0.4	6	2.50	13
Assistance to other child	0.1	1	2.00	2
Other committed work	1.0	14	2.86	36
Travel committed work	0.3	4	3.00	11
Private needs	7.1	102	1.63	149
Social life	4.8	69	1.70	105
Active leisure	8.2	118	2.64	278
Passive leisure	9.2	133	1.59	189
Non-work travel	1.2	17	2.76	42
Sleep	40.5	583	1.00	521
Totals:	100.0	1440	—	2215

Note: Data apply to a school day.
   Body weight: 36.3 kg
   BMR/day: 1286 kcal or 0.893 kcal/min
Energy expenditure = BMR/min × Time min (Column 2) × Energy Index (Column 3)
Estimated BMR factor = 1.72

B. Free day at home	Duration % of day	Time min	Energy Index	Energy expenditure kcal
Column	1	2	3	
Activity group				
School duties	1.9	28	1.61	40
Travel to school	—	—	—	—
Assistance labour force	0.4	6	2.66	14
Housework	1.4	20	2.35	42
Assistance to other child	—	—	—	—
Other committed work	1.6	23	2.35	48
Travel committed work	0.2	3	2.33	6
Private needs	6.5	94	1.66	139
Social life	11.9	171	1.67	255
Active leisure	13.9	200	2.57	459
Passive leisure	15.5	223	1.60	319
Non-work travel	3.0	43	2.84	109
Sleep	43.7	629	1.00	562
Totals:	100.0	1440	—	1993

Note: Data apply to a school day.
   Body weight: 36.3 kg
   BMR/day: 1286 kcal or 0.893 kcal/min
Energy expenditure = BMR/min × Time min (Column 2) × Energy Index (Column 3)
Estimated BMR factor = 1.55.

Table 8.3 Observed physical activity patterns expressed as BMR factors in 10–17+-year-old adolescents studied in Finland[3]

Age	Males				Females			
	At school	Free day	Daily work average	Average PAL proposed by FAO	At school	Free day	Daily work average	Average PAL proposed by FAO
10–11+	1.72	1.55	1.67	1.75	1.59	1.48	1.58	1.64
12–14+	1.68	1.58	1.65	1.67	1.67	1.52	1.63	1.58
15–17+	1.63	1.54	1.60	1.61	1.55	1.50	1.54	1.53

Table 8.4 Energy cost of observed activity patterns of elderly groups of men and women aged 60–64+ years living in Finland[3]

	BMR factor	
	Men	Women
1. Days when some work is undertaken	1.62	1.62
2. Days without work	1.52	1.55

Note: Activity patterns in Finland were monitored for 2 days each as part of an economic survey.

Table 8.5 Time allocation (as per cent of 24 hours) of elderly men and women living in Italy[4]

Age groups (years)	Women			Men		
	60–69+	70–79+	> 79	60–69+	70–79+	> 79
Activity						
Sleep	31	33	34	31	35	35
Rest in bed	8	10	13	8	8	12
Sitting						
Inactive	22	24	27	25	27	28
Active	8	9	8	8	8	8
Standing						
Inactive	4	3	3	6	6	4
Active	4	3	1	12	10	8
Walk, stairs	3	3	2	6	6	4
Moving about	19	15	11	12	10	8
Other activity	<1	<1	<1	<1	<1	<1
	100	100	100	100	100	100

men and 1.28 for the women when assessed from energy intake. The values were somewhat higher at 1.69 and 1.36 when based on measures of the observed physical activity patterns.[4]

8.3 Requirements for individuals

All the calculations in this manual apply to groups and not to individuals. Yet there is a recurrent demand by nutritionists to apply the new requirement estimates to individuals. This can be done provided it is clearly understood that whereas it is possible to estimate the requirements of a group with an error of 2–5 per cent, this error increases substantially when applied to individuals. A coefficient of variation between individuals' energy needs of ± 10 per cent is reasonable but this means that there can be very substantial differences between individuals. The error cannot be reduced easily by monitoring physical activity patterns because the principal determinant of the variation is the basal metabolic rate of the individual. Table 8.6 shows how to minimize the error of estimates.

Table 8.6 Specifying individual energy requirements

1. Measure subjects' body weight.
2. Keep a record of physical activity for a week by recording activity in 15-minute intervals using a simple diary.
3. Apply values for energy cost in Appendix 4 using energy indices for each activity.
4. Estimate BMR from the equations appropriate to the sex and age of the subject (Table 1.7) and build up table of energy expenditure as in Table 1.9.
5. Improve the accuracy of estimates by measuring the BMR of the individual. On this basis the error in estimating total expenditure will be reduced from ± 10 per cent to about ± 5 per cent if activity is monitored at 15-minute intervals and estimated costs applied. If more detailed activity monitoring is arranged, i.e. on a minute-by-minute basis with some individual measurements of the energy cost of typical physical activities, then the error of the estimate of total energy expenditure will be reduced to at best ± 3 per cent. This latter error is based on comparing estimates from physical activity monitoring with measurements of total energy expenditure by the D_2O^{18} method. Obviously the error will be affected by the accuracy of activity monitoring as well as on conducting BMR measurements with experienced investigators and trained subjects.

References

[1]Rivers, J. P. W. and Seaman, J. A. (1988). *Nutritional Aspects of Emergency Food Relief.* Paper presented at a Conference on Nutrition in Times of Disaster, 27–30 Sept., 1988, WHO, Geneva.

[2]*Drought relief in Ethiopia: a practical guide. Planning and management of feeding programmes.* Compiled by J. Appleton with the SCF Ethiopia Team. Save the Children, 17 Grove Lane, London, SE5 8RD, UK

[3]Unpublished data. P. François, ESN, FAO, Rome.

[4]Ferro-Luzzi, A. (1987). *Time allocation and activity patterns of the elderly.* Background paper presented for FAO, Nat. Inst. Nutrition, Rome.

APPENDICES

APPENDICES

Appendix 1.1 Background to population data

The United Nations (UN) secretariat provides population estimates by age–sex groups for all countries, derived in most instances from available national data that have been evaluated and adjusted where necessary. If data are unavailable or inadequate, demographic *estimates* are determined in accordance with what is 'reasonable' for, and in agreement with, any existing reliable information on a particular population.

Ten assessments of world population have been produced since 1951. The last, covering 1984, was published in 1986 and contains *estimates* and *projections* for the period 1950–2015. These estimates are based on a large variety of demographic indicators relating to population composition, growth, fertility, mortality, urbanization and international migration, the values for 1985 being presented in this report.

The assessments of population change assume that orderly progress will be made, and that catastrophes such as famines and epidemics will not occur. The projections for the period 1985–2015 are given in four variants (low, medium, high, and constant), developed from combining different assumptions about levels of fertility, mortality, and international migration; the *medium* variant represents the most plausible course of trends and is used in this manual.

Population age ranges

The UN population estimates are given by sex in 5-year age bands. By means of interpolation, these data have been disaggregated into the age groups needed for the calculation of energy requirements. The grouping of data is as follows:

Age (years)	Population data needed
0– 9 +	Yearly*
10–17 +	Yearly*
18–29 +	Single group
30–59 +	Single group
> 60	Single group

*Yearly population estimates are derived from 5-yearly aggregations using a 'Sprague multiplier'[1].

Reference

[1]Shyrock, H. S., Siegel, J. S. *et al.* (1973). *The Methods and Materials of Demography, Vol. 2.* Social and Economic Statistics Administration, Department of Commerce, Washington, D.C.

Countries	Sex	0 yrs	1 yr	2 yrs	3 yrs	4 yrs	5 yrs	6 yrs	7 yrs	8 yrs	9 yrs	10 yrs
Africa												
Algeria	M	426	407	390	374	360	348	337	327	318	309	301
	F	399	385	372	359	348	337	327	318	309	300	292
Angola	M	174	165	157	150	143	137	131	126	121	117	113
	F	173	165	157	150	144	138	132	127	122	118	114
Benin	M	87	83	79	75	71	68	64	61	58	55	52
	F	88	83	79	74	71	67	64	61	59	56	54
Botswana	M	25	24	23	22	21	20	19	18	17	16	15
	F	24	23	22	21	20	19	18	18	17	16	15
Burkina Faso	M	141	131	122	115	109	103	99	96	93	91	89
	F	140	130	122	115	109	104	100	97	94	91	89
Burundi	M	96	92	88	84	80	76	73	69	66	62	59
	F	96	91	87	83	79	76	72	69	66	63	60
Cameroon	M	189	181	173	166	159	152	146	140	134	129	124
	F	187	179	172	165	158	151	145	140	134	129	124
Cape Verde	M	4	4	5	5	5	5	5	5	5	5	4
	F	3	4	5	5	5	5	5	5	5	5	4
C.A.R.	M	48	46	45	43	41	39	37	36	34	33	31
	F	49	46	44	42	41	39	37	36	34	33	32
Chad	M	96	91	85	81	77	73	70	68	65	63	61
	F	96	90	85	81	77	74	71	68	66	63	62
Comoros	M	9	9	8	8	8	7	7	7	6	6	6
	F	9	9	8	8	8	7	7	7	6	6	6
Congo	M	34	32	31	29	28	26	25	24	24	23	22
	F	34	32	30	29	28	26	25	24	24	23	22
Cote d'Ivoire	M	200	192	184	176	168	161	154	147	140	134	128
	F	198	190	182	174	167	159	152	146	139	133	127
Equatorial Guinea	M	7	7	6	6	6	6	5	5	5	5	5
	F	7	7	6	6	6	6	5	5	5	5	5
Ethiopia	M	920	860	806	758	716	680	647	619	595	573	555
	F	928	863	806	756	712	674	641	612	588	566	548
Gabon	M	18	17	16	15	14	13	13	12	12	12	12
	F	18	17	16	15	14	13	13	12	12	12	12
Gambia	M	13	12	11	10	10	9	9	9	8	8	8
	F	13	12	11	11	10	10	9	9	8	8	8
Ghana	M	286	272	259	247	235	225	215	206	198	190	182
	F	282	269	257	245	234	224	214	205	197	189	182
Guinea	M	120	113	107	101	96	91	87	83	80	77	74
	F	120	113	107	101	96	92	88	84	81	78	75
Guinea-Bissau	M	14	14	14	14	14	13	13	12	12	11	10
	F	14	14	14	14	14	13	13	12	12	11	11
Kenya	M	508	481	455	431	409	388	369	351	334	318	303
	F	501	475	450	428	406	386	367	350	333	317	303
Lesotho	M	28	27	25	24	23	22	21	21	20	19	19
	F	28	27	26	25	24	23	22	21	20	20	19
Liberia	M	46	44	43	41	40	38	36	34	32	30	29
	F	45	44	42	41	39	37	36	34	32	31	29
Madagascar	M	196	189	181	174	167	160	153	146	140	134	128
	F	194	185	177	169	162	155	149	143	137	132	127
Malawi	M	155	143	133	124	116	109	104	99	96	93	90
	F	155	142	130	120	113	106	101	98	95	93	91
Mali	M	175	164	154	145	137	130	124	118	114	110	106
	F	174	163	154	146	138	131	125	119	114	110	106
Mauritania	M	41	38	36	34	32	31	29	28	27	26	25
	F	41	38	36	34	32	31	29	28	27	26	25

11 yrs	12 yrs	13 yrs	14 yrs	15 yrs	16 yrs	17 yrs	18–29+ yrs	30–59+ yrs	> 60 yrs	Total pop.	Number of births/1000
294	287	279	271	263	256	248	2293	2211	527	21 718	862
284	276	269	261	254	247	239	2207	2476	634		
109	105	101	97	93	90	87	840	1054	195	8754	389
110	106	102	98	95	91	88	864	1114	239		
50	48	46	44	43	42	41	380	457	86	4050	191
52	50	48	46	45	43	42	397	482	100		
15	14	14	13	13	13	12	111	96	15	1107	50
15	14	13	13	13	12	12	115	129	22		
88	86	83	80	77	74	71	682	857	150	6942	313
87	85	83	80	77	74	72	683	891	182		
56	53	51	50	48	47	45	434	566	111	4721	208
57	55	53	51	50	48	47	453	615	143		
119	114	110	106	102	99	96	951	1200	272	9873	396
119	115	111	107	103	99	96	971	1273	334		
4	4	4	4	4	4	4	39	23	10	326	9
4	4	4	4	4	4	5	44	37	13		
30	28	27	26	26	25	24	243	318	68	2576	109
31	30	29	28	27	26	25	258	350	89		
60	58	57	55	53	52	50	497	629	130	5018	210
60	58	57	55	53	52	50	504	666	159		
6	6	5	5	5	5	4	43	51	9	444	19
6	6	5	5	5	5	5	44	53	11		
21	21	20	19	19	18	17	168	213	43	1740	73
22	21	20	19	19	18	17	172	225	52		
122	117	112	108	104	100	97	956	1281	233	9810	410
122	117	112	109	105	102	99	898	1134	231		
5	4	4	4	4	4	4	38	50	12	392	16
5	4	4	4	4	4	4	39	54	14		
539	524	508	491	477	463	450	4365	5243	873	43 557	2040
533	519	503	486	472	458	445	4358	5396	1029		
11	11	11	11	11	10	10	108	178	49	1151	37
11	11	11	11	11	11	10	110	186	59		
8	7	7	7	7	7	6	64	82	15	643	30
8	7	7	7	7	7	6	66	86	18		
176	169	163	158	152	147	142	1319	1505	282	13 588	589
175	169	163	157	152	148	143	1340	1580	336		
72	70	68	66	65	64	62	598	774	136	6075	269
73	71	69	67	66	64	63	613	795	160		
10	9	9	9	9	8	8	82	119	27	889	34
10	10	10	9	9	9	9	89	126	33		
289	276	263	250	238	227	216	1934	1932	278	20 600	1029
289	276	263	250	238	227	216	1941	1998	333		
18	18	17	17	16	16	15	145	185	36	1520	60
18	18	17	17	16	16	15	155	212	50		
27	25	24	23	23	22	21	204	254	50	2191	99
27	26	25	24	23	22	22	208	260	57		
122	116	112	108	104	101	98	942	1239	257	10 012	415
123	119	115	111	107	104	100	981	1265	291		
88	86	83	80	77	75	72	667	777	130	6944	343
91	89	87	84	81	78	76	707	849	161		
102	99	96	93	90	88	85	737	884	159	8082	382
103	99	96	93	90	87	84	816	1018	206		
24	23	22	21	20	20	19	184	215	39	1888	88
24	23	22	21	21	20	19	187	222	47		

Countries	Sex	0 yrs	1 yr	2 yrs	3 yrs	4 yrs	5 yrs	6 yrs	7 yrs	8 yrs	9 yrs	10 yrs
Mauritius	M	12	12	13	13	12	12	12	12	11	11	10
	F	11	12	12	12	12	12	12	11	11	10	10
Morocco	M	387	356	332	314	301	292	287	285	285	287	290
	F	373	343	320	302	289	281	276	274	274	275	279
Mozambique	M	266	253	241	230	220	212	204	196	190	184	178
	F	260	248	237	227	218	209	202	194	188	181	176
Namibia	M	31	29	28	27	25	24	23	22	21	21	20
	F	31	29	28	26	25	24	23	22	21	21	20
Niger	M	134	126	118	111	105	100	95	91	87	84	81
	F	133	125	118	112	106	101	96	92	88	84	81
Nigeria	M	2120	2003	1895	1796	1706	1624	1548	1479	1416	1358	1305
	F	2098	1985	1882	1787	1699	1619	1546	1478	1416	1359	1307
Rwanda	M	143	131	122	113	106	100	95	91	88	85	82
	F	142	131	121	113	106	100	95	91	88	85	83
Senegal	M	128	121	115	109	104	100	96	92	89	86	83
	F	127	121	115	110	105	100	96	93	89	86	83
Sierra Leone	M	69	64	60	57	54	51	49	47	45	43	42
	F	69	65	61	58	55	52	50	48	46	44	43
Swaziland	M	14	13	12	12	11	11	10	10	9	9	9
	F	13	13	12	12	11	11	10	10	9	9	9
Tanzania, U.R. of	M	503	479	455	433	412	393	374	356	339	323	309
	F	503	475	450	427	406	388	370	355	340	327	315
Togo	M	60	57	54	51	49	47	45	43	41	39	38
	F	59	56	54	51	49	47	45	43	41	39	38
Tunisia	M	96	100	102	103	103	102	100	98	96	93	90
	F	94	96	97	97	97	96	94	92	90	88	86
Uganda	M	344	325	308	293	278	264	252	240	230	220	211
	F	340	322	306	291	277	263	251	240	229	219	210
Zaire	M	602	572	545	520	497	476	457	439	423	408	394
	F	596	567	540	516	494	474	455	438	423	408	395
Zambia	M	145	137	129	123	117	111	106	102	98	94	91
	F	142	135	128	122	116	111	106	102	97	94	90
Zimbabwe	M	190	180	172	164	156	149	142	136	131	125	120
	F	187	178	170	162	155	148	141	135	130	125	120
Asia												
Bangladesh	M	1940	1889	1838	1787	1735	1684	1633	1582	1532	1482	1433
	F	1871	1795	1726	1665	1609	1559	1513	1471	1432	1395	1360
Bhutan	M	24	23	22	21	21	20	20	19	19	18	18
	F	23	22	21	20	19	19	18	18	17	17	17
Burma	M	530	509	492	480	471	466	462	461	461	461	462
	F	512	492	477	465	457	452	449	448	448	449	450
China	M	10 877	9958	9363	9055	8995	9146	9470	9929	10 485	11 101	11 773
	F	10 158	9291	8733	8449	8401	8553	8869	9313	9849	10 440	11 083
Democratic Kampuchea	M	227	164	115	78	52	35	27	27	31	40	53
	F	220	160	113	78	53	37	30	29	34	42	55
India	M	10 644	10 396	10 187	10 014	9870	9751	9651	9565	9488	9414	9347
	F	10 064	9813	9596	9411	9253	9117	8999	8894	8798	8707	8623
Indonesia	M	2430	2351	2288	2240	2204	2179	2162	2150	2141	2133	2128
	F	2348	2268	2205	2159	2125	2103	2089	2082	2078	2076	2077
Japan	M	827	795	781	781	793	815	844	877	913	949	987
	F	780	751	738	738	751	772	800	833	868	903	939
Korea, D.P.R. of	M	299	297	294	290	286	282	278	273	268	262	257
	F	286	279	273	268	263	259	256	252	249	247	244
Korea, Rep. of	M	469	461	454	448	444	440	437	436	435	434	435
	F	432	429	425	422	419	416	414	412	411	410	409
Lao P.D.R.	M	73	70	68	65	63	61	59	58	56	55	54
	F	71	69	66	64	62	60	58	56	55	54	52
Malaysia	M	254	236	221	210	201	195	191	188	187	187	188
	F	241	223	209	198	190	184	181	179	178	178	180

11 yrs	12 yrs	13 yrs	14 yrs	15 yrs	16 yrs	17 yrs	18–29+ yrs	30–59+ yrs	> 60 yrs	Total pop.	Number of births/1000
10	10	10	10	11	11	11	138	154	25	1050	25
9	9	9	10	10	11	11	135	166	35		
295	296	292	284	277	270	261	2402	2549	636	21 941	752
284	285	280	272	264	257	248	2434	2744	607		
173	168	162	157	152	147	142	1365	1731	339	13 961	589
170	165	160	155	151	147	142	1379	1833	408		
19	19	18	17	17	16	16	153	184	36	1550	65
19	18	18	17	17	16	16	153	195	43		
78	76	73	70	67	65	62	589	685	133	6115	291
78	75	72	69	67	65	62	589	711	165		
1256	1209	1165	1122	1082	1045	1006	9074	10 191	1727	95 198	4431
1258	1212	1168	1126	1086	1050	1011	9230	10 699	2057		
80	78	75	72	69	65	63	590	634	110	6070	291
81	79	76	72	69	66	63	604	676	138		
80	78	75	72	70	67	65	631	784	144	6444	281
80	78	75	73	70	68	66	646	807	167		
41	40	39	38	37	36	35	357	478	81	3602	163
42	40	39	38	37	37	36	365	510	102		
8	8	8	7	7	7	7	63	74	14	650	28
8	8	8	7	7	7	7	64	78	17		
294	281	270	259	249	240	230	2120	2395	382	22 499	1042
304	293	281	269	257	246	236	2172	2518	473		
37	35	34	33	32	31	30	287	348	68	2960	124
37	35	34	33	32	31	31	293	368	82		
86	84	84	85	86	87	87	803	820	255	7081	221
83	81	81	81	81	81	80	803	905	218		
202	194	187	180	174	168	162	1498	1645	293	15 477	718
202	194	187	181	175	169	164	1523	1724	340		
382	369	356	343	330	318	306	2946	3459	592	29 938	1259
383	370	356	342	328	315	302	2920	3772	809		
88	85	82	78	75	72	69	646	734	131	6666	296
86	83	80	77	74	71	68	647	766	157		
115	111	106	102	98	95	91	850	948	176	8777	380
115	110	106	102	98	95	91	858	988	207		
1384	1337	1292	1247	1204	1162	1121	10 711	11 534	2609	101 147	4240
1328	1292	1247	1197	1150	1103	1059	10 137	10 877	2223		
18	18	17	17	16	16	15	152	202	37	1417	52
16	16	16	15	15	14	14	139	191	40		
465	463	455	441	429	416	403	4073	5078	1136	37 153	1079
453	451	443	430	418	406	393	4001	5095	1347		
12 498	13 065	13 365	13 466	13 562	13 639	13 542	126 496	175 297	40 713	1 059 521	19 532
11 776	12 314	12 594	12 681	12 764	12 829	12 723	117 474	159 273	46 158		
71	85	90	91	93	96	97	955	1046	150	7284	311
71	84	89	89	91	93	93	923	1093	185		
9289	9194	9040	8847	8657	8462	8268	84 753	111 672	26 152	758 927	22 962
8547	8444	8294	8114	7937	7756	7577	77 320	105 319	25 683		
2127	2109	2064	2002	1941	1880	1818	17 881	22 224	4439	166 440	5101
2082	2070	2030	1972	1916	1859	1801	18 011	23 241	4957		
1030	1050	1038	1003	970	934	904	9947	25 843	7353	120 742	1558
980	1000	988	954	922	888	859	9628	26 086	10 128		
251	246	241	237	233	229	224	2272	2609	474	20 385	586
241	238	234	230	226	221	217	2334	2774	692		
436	438	440	442	444	445	449	5404	6314	1108	41 258	915
409	409	411	413	414	415	418	5029	6371	1557		
52	51	50	49	48	46	45	415	536	98	4117	159
51	50	49	48	46	45	44	406	528	110		
190	190	188	183	179	175	171	1761	2112	423	15 557	453
183	183	181	176	171	167	163	1800	2092	466		

Countries	Sex	0 yrs	1 yr	2 yrs	3 yrs	4 yrs	5 yrs	6 yrs	7 yrs	8 yrs	9 yrs	10 yrs
Mongolia	M	32	31	30	29	29	28	27	27	26	25	25
	F	31	30	29	29	28	27	26	26	25	25	24
Nepal	M	297	286	276	267	260	254	248	244	239	234	230
	F	280	269	259	251	243	237	231	225	221	216	212
Pakistan	M	1988	1886	1794	1711	1637	1570	1511	1459	1413	1372	1336
	F	1870	1768	1676	1593	1518	1452	1394	1342	1297	1258	1223
Philippines	M	814	820	821	818	811	800	786	770	752	733	713
	F	772	780	782	780	773	762	749	733	715	696	675
Sri Lanka	M	226	221	215	209	203	198	192	187	182	178	173
	F	218	212	206	200	194	188	183	179	175	171	168
Thailand	M	713	675	645	625	611	604	602	604	610	617	627
	F	686	648	619	598	585	578	576	579	585	592	602
Vietnam	M	759	800	828	844	849	847	837	822	802	781	756
	F	655	749	817	861	885	891	882	861	832	797	755
Europe												
Albania	M	39	39	39	39	39	38	38	37	37	36	36
	F	37	37	37	37	37	36	36	35	35	34	34
Austria	M	52	49	47	45	44	44	44	44	45	46	47
	F	49	47	44	43	42	41	41	42	42	44	45
Belgium	M	65	64	62	62	61	61	61	62	63	63	65
	F	62	60	59	59	58	58	59	59	60	61	62
Bulgaria	M	72	71	71	70	70	70	69	69	69	69	68
	F	68	67	67	67	66	66	66	66	66	65	65
Cyprus	M	7	7	7	7	7	7	6	6	6	5	5
	F	6	6	7	7	6	6	6	6	5	5	5
Czechoslovakia	M	103	113	121	128	133	136	139	140	140	139	137
	F	99	108	116	122	127	130	133	134	134	133	131
Denmark	M	27	28	28	29	29	30	31	32	33	34	35
	F	26	26	27	27	28	29	30	31	32	33	34
Finland	M	31	32	33	34	34	34	33	33	32	32	31
	F	29	31	32	33	33	32	32	31	31	30	29
France	M	432	406	387	375	368	366	369	374	382	391	402
	F	421	394	374	361	354	351	352	357	364	373	384
Germany, F.R. of	M	323	318	312	307	303	299	298	298	300	305	311
	F	308	302	296	291	287	284	283	283	286	292	298
Greece	M	69	71	73	74	75	75	75	74	74	73	72
	F	68	69	70	70	70	70	70	69	69	68	68
Hungary	M	53	63	71	77	81	85	87	88	88	87	85
	F	52	61	68	73	77	80	82	83	83	82	81
Iceland	M	3	2	2	2	2	2	2	2	2	2	2
	F	2	2	2	2	2	2	2	2	2	2	2
Ireland	M	40	39	38	37	36	36	35	35	35	36	36
	F	38	37	35	35	34	34	34	34	34	34	34
Israel	M	52	50	48	46	45	45	45	45	45	45	45
	F	50	47	45	44	43	42	42	42	42	43	43
Italy	M	315	315	319	326	335	347	360	374	389	404	419
	F	302	302	305	311	320	330	342	355	369	383	397
Luxembourg	M	2	2	2	2	2	2	2	2	2	2	2
	F	2	2	2	2	2	2	2	2	2	2	2
Malta	M	3	3	3	3	3	3	3	3	3	3	3
	F	3	3	3	3	3	3	3	3	3	3	3
Netherlands	M	97	93	89	88	87	87	89	91	93	97	100
	F	95	90	87	84	83	84	85	86	89	92	96
Norway	M	28	27	26	25	25	25	26	27	28	29	30
	F	28	26	25	24	24	24	25	25	26	27	29
Poland	M	328	336	341	343	343	341	337	332	325	318	310
	F	311	319	324	327	327	326	322	317	311	304	296
Portugal	M	86	86	86	86	86	85	85	85	85	85	85
	F	79	80	82	82	83	83	83	83	83	83	82

11 yrs	12 yrs	13 yrs	14 yrs	15 yrs	16 yrs	17 yrs	18–29+ yrs	30–59+ yrs	>60 yrs	Total pop.	Number of births/1000
24	24	23	23	22	22	21	198	243	47	1908	64
23	23	23	22	22	22	22	197	244	53		
227	221	212	201	190	179	170	1689	2123	404	16 482	649
208	203	196	187	179	171	163	1548	2119	413		
1303	1275	1253	1233	1216	1200	1178	11 328	12 058	2358	100 380	4013
1191	1165	1147	1132	1119	1109	1091	10 452	11 326	2178		
690	670	656	645	633	622	608	5863	6965	1364	54 498	1709
653	635	621	611	601	590	580	6141	6992	1500		
169	166	166	167	169	170	171	1804	2429	596	16 205	439
164	162	162	162	163	164	164	1823	2409	548		
638	646	646	642	638	633	624	6111	6967	1320	51 411	1369
614	622	622	617	613	608	599	5982	7104	1586		
726	707	707	718	726	733	735	7290	6127	1640	59 713	1763
704	670	665	679	688	696	700	7141	7689	2060		
35	34	34	33	33	32	32	358	434	100	3050	80
33	32	32	32	31	31	31	344	429	113		
49	51	53	55	58	60	63	748	1387	539	7502	94
46	48	51	53	55	58	60	723	1412	946		
66	67	69	70	71	73	74	964	1912	789	9903	123
63	64	65	67	68	70	71	926	1898	1111		
68	68	67	66	65	64	64	765	1795	718	9071	141
65	64	64	63	62	61	60	730	1814	849		
5	4	4	5	5	5	5	72	119	41	669	13
5	4	4	4	5	5	5	68	121	49		
136	132	128	122	116	110	106	1357	2927	1036	15 579	238
130	127	122	117	111	105	102	1299	2996	1504		
37	38	38	39	40	40	41	472	994	447	5122	55
35	36	37	37	37	38	38	454	979	585		
30	30	31	32	34	35	37	471	992	316	4891	64
29	28	29	31	33	34	36	454	982	525		
416	426	428	426	425	423	423	5159	10 466	3998	54 621	787
398	407	410	408	407	406	406	4993	10 207	5656		
317	332	363	403	441	480	510	6122	12 483	4388	60 877	617
304	319	348	385	421	458	486	5777	12 155	7800		
71	71	73	75	78	81	82	884	1845	790	9878	146
67	67	68	71	73	76	77	842	1951	971		
83	81	80	78	77	75	74	870	2105	788	10 697	138
79	77	75	74	72	71	69	828	2199	1159		
2	2	2	2	2	2	2	27	41	15	243	4
2	2	2	2	2	2	2	26	39	18		
36	36	36	36	36	35	35	368	556	236	3608	75
35	35	35	34	34	33	33	354	537	284		
46	46	45	44	43	41	40	418	647	240	4252	94
44	44	43	42	41	40	38	396	676	286		
435	447	455	459	462	464	466	5356	11 011	4553	57 300	640
412	424	432	436	440	443	445	5152	11 258	6132		
2	2	2	2	3	3	3	34	77	26	363	4
2	2	2	2	2	3	3	34	74	39		
3	3	3	3	3	2	2	39	68	23	383	6
3	3	3	3	3	3	3	37	80	30		
104	108	112	116	120	124	127	1522	2831	1028	14 500	179
100	104	108	111	115	119	122	1454	2730	1365		
31	32	33	34	34	35	35	389	749	382	4142	51
30	31	32	32	32	33	33	370	726	491		
302	293	283	274	265	256	252	3622	6913	2029	37 187	674
288	280	271	262	254	246	241	3461	7145	3109		
84	85	85	86	87	87	88	1046	1618	643	10 212	172
82	82	82	83	83	84	85	1027	1943	910		

Countries	Sex	0 yrs	1 yr	2 yrs	3 yrs	4 yrs	5 yrs	6 yrs	7 yrs	8 yrs	9 yrs	10 yrs
Romania	M	183	192	198	203	205	206	206	205	203	200	197
	F	172	181	188	192	195	197	196	195	194	191	188
Spain	M	221	257	286	308	325	337	344	348	350	348	345
	F	218	246	270	289	303	314	321	326	328	328	326
Sweden	M	51	49	48	47	47	48	48	50	51	52	54
	F	50	47	46	45	45	46	46	47	49	50	52
Switzerland	M	40	38	37	36	36	35	36	36	37	38	39
	F	37	36	35	34	34	34	34	35	35	36	37
Turkey	M	710	670	640	617	601	591	586	584	586	590	596
	F	692	650	617	593	575	564	558	555	556	559	564
United Kingdom	M	436	401	375	356	343	337	336	339	347	357	370
	F	415	381	356	338	326	320	318	321	328	338	350
Yugoslavia	M	172	181	188	193	196	198	198	197	195	193	190
	F	159	169	176	182	185	187	187	186	185	182	179

Latin America and the Caribbean

Countries	Sex	0 yrs	1 yr	2 yrs	3 yrs	4 yrs	5 yrs	6 yrs	7 yrs	8 yrs	9 yrs	10 yrs
Argentina	M	353	355	355	353	349	343	336	328	319	310	300
	F	341	344	344	342	338	333	326	318	310	301	291
Barbados	M	3	2	2	2	2	2	2	2	2	2	2
	F	2	2	2	2	2	2	2	2	2	2	2
Bolivia	M	122	116	111	106	102	98	94	91	88	85	82
	F	119	114	109	105	101	97	94	91	88	85	82
Brazil	M	1900	1861	1823	1785	1748	1711	1676	1642	1609	1577	1547
	F	1857	1824	1791	1758	1725	1693	1661	1630	1600	1570	1542
Chile	M	146	136	128	122	118	116	115	116	117	118	121
	F	140	130	122	117	113	111	111	111	112	114	117
Colombia	M	414	408	401	393	384	375	366	357	349	341	332
	F	400	395	389	382	374	366	357	348	340	332	324
Costa Rica	M	38	38	37	37	36	35	34	32	31	30	29
	F	37	36	36	35	34	33	32	31	30	29	28
Cuba	M	102	88	79	73	70	71	73	77	83	89	97
	F	98	85	75	70	67	67	70	74	79	86	93
Dominican Republic	M	103	96	90	87	84	82	81	81	82	82	83
	F	99	92	87	84	81	80	79	79	80	80	82
Ecuador	M	162	155	150	145	141	137	133	130	127	124	122
	F	156	150	145	140	136	132	129	126	123	121	119
El Salvador	M	103	100	97	95	92	89	86	83	81	78	75
	F	99	96	94	91	89	86	83	81	78	76	73
Guatemala	M	155	151	146	142	137	133	128	123	119	115	110
	F	149	145	141	136	132	128	123	119	115	111	107
Guyana	M	12	13	13	13	13	13	12	12	12	12	11
	F	12	12	12	12	12	12	12	12	12	11	11
Haiti	M	122	117	112	108	104	101	97	94	92	89	86
	F	119	114	110	106	103	99	96	93	90	88	86
Honduras	M	86	83	81	78	76	73	71	68	66	64	61
	F	84	82	79	77	75	72	70	68	66	63	61
Jamaica	M	34	32	31	30	29	28	28	28	28	28	28
	F	33	31	30	29	28	27	27	26	26	26	27
Mexico	M	1255	1220	1192	1170	1153	1140	1130	1122	1115	1108	1102
	F	1205	1173	1147	1127	1111	1100	1091	1083	1077	1070	1065
Nicaragua	M	67	64	61	59	56	54	52	51	49	47	46
	F	65	62	59	57	55	53	51	49	47	46	45
Panama	M	30	29	29	28	28	28	28	28	28	28	28
	F	29	28	27	27	27	27	26	26	26	26	26
Paraguay	M	63	61	59	58	56	55	53	51	50	48	47
	F	61	59	58	56	55	53	52	50	49	47	46
Peru	M	329	316	304	294	285	277	270	264	258	253	249
	F	317	304	293	284	275	268	261	255	250	246	241
Suriname	M	6	6	5	5	4	4	4	4	4	4	5
	F	6	5	5	4	4	4	4	4	4	4	4

11 yrs	12 yrs	13 yrs	14 yrs	15 yrs	16 yrs	17 yrs	18–29+ yrs	30–59+ yrs	> 60 yrs	Total pop.	Number of births/1000
192	190	194	201	207	215	215	1999	4355	1407	23 017	392
183	181	185	192	199	207	207	1924	4414	1865		
338	334	334	337	338	338	338	3726	6832	2578	38 542	553
322	320	320	322	322	322	321	3607	6977	3479		
56	58	59	60	60	61	61	710	1610	846	8351	94
53	55	56	57	58	59	60	677	1572	1055		
40	42	43	45	47	48	50	589	1280	514	6374	73
38	39	41	43	45	46	48	578	1295	710		
604	607	603	594	585	576	566	5719	7225	1499	49 289	1416
571	574	566	552	539	526	514	5259	6704	1654		
385	400	413	426	438	451	460	5294	10 308	4834	56 125	760
364	378	390	400	411	421	430	5089	10 244	6802		
186	184	183	183	183	184	184	2269	4562	1216	23 153	372
176	173	173	174	175	176	176	2177	4627	1712		
290	280	272	265	259	252	246	2800	5111	1686	30 564	722
282	272	265	258	252	245	240	2733	5146	2120		
3	3	3	3	3	3	2	30	34	14	253	4
3	3	3	3	3	2	2	31	40	21		
79	77	74	72	69	67	65	622	766	152	6371	263
80	77	75	73	70	68	66	643	817	178		
1519	1493	1469	1447	1426	1402	1388	15 444	18 935	4277	135 564	3929
1515	1489	1464	1441	1418	1393	1377	15 382	19 109	4648		
124	126	126	125	123	122	121	1430	1880	431	12 038	262
120	122	122	121	120	118	118	1409	1960	569		
324	319	320	323	327	330	332	3420	3823	769	28 714	845
316	311	312	316	320	324	325	3369	3782	925		
28	27	27	28	28	29	29	321	347	71	2600	74
27	26	26	27	27	28	28	311	345	81		
106	112	115	116	117	117	118	1178	1667	563	10 038	167
101	107	110	111	112	112	112	1122	1627	547		
85	85	84	82	80	78	75	729	744	147	6243	195
83	84	82	80	78	76	74	726	747	146		
120	117	115	113	110	108	105	1013	1147	244	9378	322
116	114	112	110	108	105	103	993	1150	272		
73	71	68	66	64	62	60	586	630	128	5552	208
71	69	67	65	63	61	59	575	631	157		
106	102	99	95	92	89	86	800	914	185	7963	317
103	99	96	92	89	86	83	783	909	192		
11	11	11	11	11	10	10	123	117	27	953	26
11	11	10	10	10	10	10	120	122	30		
84	82	80	77	75	73	70	670	754	160	6585	256
83	81	79	76	74	72	70	677	827	193		
59	57	55	53	51	50	48	441	475	97	4372	177
59	57	55	53	51	49	47	436	473	103		
28	28	29	29	29	29	30	310	232	91	2336	63
27	27	28	28	28	28	29	303	261	110		
1097	1084	1057	1020	985	948	913	8662	9153	1930	78 996	2516
1060	1047	1022	988	954	920	886	8511	9501	2302		
45	43	42	40	39	37	36	337	347	62	3272	133
43	42	41	39	38	37	36	339	362	72		
28	27	27	27	26	26	25	249	295	73	2180	58
27	26	26	26	25	25	24	241	281	71		
46	45	44	43	42	41	40	405	440	92	3681	123
44	43	42	42	41	41	40	404	453	108		
245	241	236	231	226	220	215	2149	2551	510	19 698	679
238	234	229	224	219	214	209	2092	2535	586		
5	5	5	5	5	5	5	45	44	12	375	10
5	5	5	5	5	5	5	46	48	13		

Countries	Sex	0 yrs	1 yr	2 yrs	3 yrs	4 yrs	5 yrs	6 yrs	7 yrs	8 yrs	9 yrs	10 yrs
Trinidad & Tobago	M	15	15	14	14	14	13	13	13	13	13	13
	F	14	14	14	14	13	13	13	13	12	12	12
Uruguay	M	29	28	28	28	28	27	27	27	27	27	27
	F	28	27	27	27	27	27	27	27	27	26	26
Venezuela	M	271	267	261	256	250	244	238	232	226	220	214
	F	260	256	251	246	240	235	229	223	217	211	206
Near East												
Afghanistan	M	324	297	274	256	242	231	223	218	215	213	212
	F	311	284	262	244	230	220	212	207	203	201	201
Bahrain	M	7	7	6	6	5	5	5	4	4	4	4
	F	7	6	6	5	5	5	5	4	4	4	4
Egypt	M	715	733	740	737	727	710	688	662	633	603	571
	F	679	695	700	697	687	670	649	624	597	568	538
Iran (I.R. of)	M	814	778	747	719	695	674	656	640	625	612	600
	F	774	735	701	672	648	627	609	594	582	571	562
Iraq	M	316	310	303	296	287	277	267	257	246	236	225
	F	302	296	289	281	273	264	254	244	234	224	214
Jordan	M	78	72	68	65	62	59	57	55	54	53	52
	F	76	70	66	62	58	56	53	52	50	49	48
Kuwait	M	38	35	32	29	27	25	23	22	21	21	20
	F	37	33	30	28	26	24	23	21	21	20	20
Lebanon	M	38	36	34	33	32	32	32	32	32	33	33
	F	37	35	33	32	32	31	31	32	32	32	33
Libyan A.J.	M	75	71	68	65	62	59	57	55	53	52	50
	F	72	69	65	62	60	57	55	53	52	50	48
Oman	M	26	25	24	22	21	20	19	18	18	17	16
	F	25	24	23	22	21	20	19	18	17	16	15
Qatar	M	7	6	5	5	4	4	3	3	3	3	3
	F	7	6	5	4	4	3	3	3	3	3	3
Saudi Arabia	M	217	215	212	207	201	194	186	178	169	160	151
	F	209	208	205	200	195	188	180	172	164	155	146
Somalia	M	90	88	86	83	80	76	73	69	66	62	59
	F	90	88	86	83	80	77	73	70	66	63	59
Sudan	M	433	415	398	381	366	351	336	323	310	297	286
	F	420	403	387	371	356	342	328	314	301	289	277
Syrian A.R.	M	229	219	210	201	193	185	177	169	162	155	148
	F	225	214	203	193	185	176	169	162	156	150	144
U.A. Emirates	M	19	19	19	18	17	16	16	15	14	13	12
	F	19	18	18	17	16	15	15	14	13	12	11
Yeman, Arab Rep.	M	141	135	129	124	119	115	110	106	102	99	95
	F	136	131	126	121	116	112	108	104	100	97	93
Yemen, P.D.R. of	M	46	42	38	36	33	32	31	30	29	29	29
	F	45	41	37	35	33	31	30	29	29	28	28
North America												
Canada	M	198	196	194	192	189	188	186	184	183	182	182
	F	187	185	183	181	179	178	176	175	174	173	172
U.S.A.	M	2065	1975	1899	1834	1782	1741	1710	1690	1678	1675	1680
	F	1965	1884	1813	1754	1705	1666	1637	1617	1606	1602	1607
Southwest Pacific												
Australia	M	138	130	124	120	118	117	118	119	122	124	128
	F	131	123	118	114	112	111	112	113	116	118	122
Fiji	M	10	10	10	10	10	9	9	9	9	8	8
	F	10	10	10	9	9	9	9	8	8	8	8
New Zealand	M	28	27	26	25	25	25	25	26	27	27	28
	F	27	25	25	24	24	24	24	25	25	26	27
Papua New Guinea	M	64	60	57	54	52	50	49	48	47	47	46
	F	62	57	54	51	48	46	45	44	43	43	43

11 yrs	12 yrs	13 yrs	14 yrs	15 yrs	16 yrs	17 yrs	18–29+ yrs	30–59+ yrs	> 60 yrs	Total pop.	Number of births/1000
13	13	13	12	12	12	12	145	166	44	1185	29
12	12	12	12	12	12	12	147	171	50		
27	27	27	26	25	25	24	283	510	204	3012	58
26	26	26	25	24	24	23	277	525	258		
208	203	200	198	196	194	191	1950	2304	429	17 317	533
200	196	193	191	189	187	185	1895	2271	488		
214	213	209	202	196	191	185	1721	2273	368	16 519	796
202	201	197	191	185	180	174	1621	2129	391		
4	4	4	4	4	4	3	65	99	7	432	12
4	4	4	4	4	3	3	43	45	47		
536	511	501	501	501	502	499	5217	6179	1332	46 909	1618
504	480	470	469	468	468	464	4875	6243	1567		
590	578	562	544	527	510	494	4850	5426	1100	44 632	1699
554	545	533	518	504	491	477	4711	5303	1182		
214	204	196	189	182	175	169	1592	1835	322	15 898	648
204	195	187	180	174	168	162	1522	1780	356		
51	50	49	47	46	44	43	377	356	72	3515	144
47	46	45	44	42	41	40	339	347	74		
20	20	19	18	18	17	17	224	381	26	1811	62
19	19	18	17	17	16	16	146	191	19		
34	34	35	35	35	34	34	264	320	97	2668	78
33	34	34	34	33	33	33	313	361	111		
49	47	45	43	41	40	38	346	511	73	3605	150
47	46	44	42	40	38	36	322	381	67		
15	14	13	13	12	12	11	127	190	25	1242	52
14	13	13	12	12	11	11	112	143	26		
3	3	3	3	3	3	3	48	83	5	315	10
3	3	3	2	2	2	2	23	29	3		
142	134	128	123	118	114	111	1355	1736	238	11 542	441
137	129	123	118	114	109	105	969	1177	250		
55	52	51	49	48	47	46	447	563	106	4653	208
56	53	51	50	49	47	46	453	588	127		
274	264	254	246	237	229	222	2147	2589	456	21 550	923
266	256	246	238	230	223	216	2113	2636	523		
142	136	130	125	119	114	111	1145	1044	218	10 505	448
139	134	128	121	115	109	104	1058	1057	231		
11	10	10	10	10	10	10	202	438	19	1327	34
10	9	9	8	7	7	6	78	105	12		
92	89	86	82	79	75	72	578	637	178	6848	311
90	87	84	81	78	75	72	676	919	200		
29	29	28	27	26	25	24	214	232	46	2137	94
28	28	28	26	25	24	23	218	263	53		
182	183	184	186	188	189	194	2789	4785	1646	25 426	375
173	173	174	176	178	180	184	2730	4775	2118		
1694	1712	1733	1759	1788	1817	1860	25 714	41 858	16 266	238 020	3718
1619	1636	1658	1685	1715	1746	1790	25 153	43 709	22 520		
132	135	136	136	137	137	137	1626	2899	1002	15 698	241
126	129	130	130	131	131	132	1567	2820	1275		
8	7	7	7	7	7	7	81	95	19	691	20
7	7	7	7	7	7	7	80	98	19		
29	30	31	31	31	31	31	361	574	213	3318	51
28	29	29	29	30	30	30	347	568	272		
46	46	45	44	43	43	42	382	480	84	3511	128
43	42	42	41	40	39	37	326	459	78		

The growth curves provided in Appendix 2.2 are *not* newly developed local standards, but simply currently available data from single studies made within some of the listed countries. The data sets vary in size and quality; some are the result of national surveys and others are taken from surveys on smaller communities within a country. Sampling techniques vary, and in many cross-sectional surveys, sample sizes have changed from year to year, thus affecting the consistency of the growth curves which is shown by wide fluctuations in percentile values between age bands. For comparative purposes, and for use in contexts where no local data are available, the curves have been modified as described below. They therefore can only be considered as 'best estimates' rather than statistically representative national data sets. Hence it is *recommended* that, where possible, local data should be used rather than values provided in the following pages.

(I) 0–17 + Years

1. Weight and height data for groups *aged 0–17 + years* have been used from a variety of sources, currently gathered together by the Food Policy and Nutrition Division, FAO.

2. For comparative purposes the weight and height curves have been smoothed, matched with the NCHS standards, and expressed as percentiles. To prevent bias, all measurements were allocated to the nearest main percentile (i.e. 3rd, 5th, 10, 20, 30, 40, 50, 60, 70, 80, 90, 95 and 97th percentiles).

3. Thus a series of 62 modified curves has been established which is provided in Appendix 2.3.

(II) Adult data

1. Complete growth curves covering the whole life span are available for only a few countries. Therefore some established characteristics of growth and anthropometry had to be used when estimating appropriate adult values. They are:

   (a) Females are regarded as having reached their maximum growth potential by 18 years.

   (b) In well nourished male populations full growth *may* be achieved by 18 years, but in less well nourished populations, growth may continue for another 4–5 years so, in the absence of data, heights must be derived.

   (c) A commonly observed feature of the relationship between male and female height is that in many populations females are approximately 7 per cent shorter than males. This relationship was therefore used to obtain adult male heights.

   (d) The body mass index (BMI), which expresses a relationship between weight and height (Wt/Ht^2) can be used to calculate an actual desirable weight from height.

2. *Height.* Where adult measurements are unavailable, the actual heights of females at 18 years has been treated as the adult height. Male heights have been estimated by calculating a value 7 per cent higher than that of the females.

3. *Weight.* Similarly, weights of females at 18 years have been treated as adult weights. Male weights have been calculated using BMIs and the estimated heights and then applying an estimate of the BMI.

   Source of male BMIs:

   (i) Studies on adults from the country itself; or

(ii) In the absence of a study, appropriate BMIs from a nearby country have been applied; or

(iii) Where no data are available for LDCs, a BMI within the range 19–21 has been selected. This range was found to apply to the LDC adult data provided by Eveleth and Tanner[1].

Note: In many cases, sources of data for the group 0–17+ and adult values are different, and often not nationally representative. This has resulted in some unsatisfactory increments between the adolescent and adult values. *Up-to-date local data should be used for the calculations wherever possible.*

4. *Patterns of weight change.* Lean body mass does not in general increase over the age of 24 years, but total body weight does, with a consequent increase in BMI. This process generally occurs in western societies and in the urban populations of some LDCs. Evidence from studies in the USA, the UK and Belgium suggest that an increment of 2 BMI points could be added to adult weights in the 30–59 years age group in order to allow for the extra energy required to maintain the actual body weight.

Reference

[1]Eveleth, P. B. and Tanner, J. M. (1984). *Worldwide variation in human growth.* International Biological Programme 8, Cambridge University Press, Cambridge.

Appendix 2.2 Index to growth curves

FAO member countries are listed by region. Adjacent numbers indicate a column in the following table, Appendix 2.3, which contains weight and height data to be used as a 'best estimate' when no local data are available.

Africa	
Algeria	55
Angola	3
Benin	3
Botswana	3
Burkino Faso	4
†Burundi	1
Camaroon	6
Cape Verde	4
C.A.R.	6
Chad	8
Comoros	11
Congo	6
*Cote d'Ivoire	3
Equatorial Guinea	8
*Ethiopia	2
Gabon	3
Gambia	4
Ghana	3
Guinea	3
Guinea-Bissau	3
Kenya	11
Lesotho	11
*Liberia	4
Madagascar	6
Malawi	11
Mali	8
Mauritania	8
†Mauritius	5
Morocco	54
*Mozambique	6
Namibia	11
Niger	8
Nigeria	4
†Rwanda	7
*Senegal	8
Sierra Leone	4
Swaziland	11
*Tanzania, U.R. of	11
Togo	3
*Tunisia	54
Uganda	11
Zaire	6
Zambia	3
Zimbabwe	3

Latin America and the Caribbean	
*Argentina	36
Barbados	45
*Bolivia	37
*Brazil	38
*Chile	39
*Colombia	40
*Costa Rica	41
*Cuba	42
Dominican Republic	42
Ecuador	39
El Salvador	43
*Guatemala	43
†Guyana	20
*Haiti	44
Honduras	43
*Jamaica	45
*Mexico	46
Nicaragua	43
*Panama	47
Paraguay	36
Peru	39
Suriname	48
†Trinidad and Tobago	48
*Uruguay	49
*Venezuela	50

Asia	
*Bangladesh	12
Bhutan	14
*Burma	13
*China	14
Democratic Kampuchea	18
*India	15
*Indonesia	16
*Japan	17
Korea, D.P.R. of	18
*Korea, Republic of	18
Lao P.D.R.	18
†Malaysia	19
Mongolia	14
Nepal	14
Pakistan	15
*Philippines	21
Sri Lanka	15
*Thailand	22
Vietnam	22

Europe	
Albania	24
Austria	29
*Belgium	23
*Bulgaria	24
Cyprus	24
*Czechoslovakia	25
Denmark	32
*Finland	26
*France	27
*Germany, F.R. of	28
Greece	24
*Hungary	29
Iceland	32
Ireland	35
*Israel	30
*Italy	62

Luxembourg	23	*Lebanon	53
Malta	24	Libyan A.J.	51
*Netherlands	31	Oman	54
*Norway	32	Qatar	54
*Poland	33	Saudi Arabia	51
Portugal	24	*Somalia	9
Romania	29	*Sudan	10
Spain	24	Syrian A.R.	53
Sweden	57	United Arab Emirates	54
*Switzerland	34	Yemen, Arab Rep	54
*Turkey	55	Yemen, P.D.R. of	54
*United Kingdom	35		
Yugoslavia	29	**North America**	
		*Canada	56
Near East		*U.S.A.	57
Afghanistan	15		
Bahrain	52	**Southwest Pacific**	
*Egypt	51	*Australia	58
Iran (I.R. of)	52	†Fiji	59
Iraq	52	*New Zealand	60
*Jordan	52	*Papua New Guinea	61
Kuwait	51		

* Original data.

† Combined data from more than one study.

Appendix 2.3 Growth curves providing weights and heights

Males: weight kg

Age yrs	1	2	3	4	5	6	7	8	9	10	11	12	13	14	15	16
0+	6.3	6.0	7.3	7.8	7.8	7.2	6.3	7.3	6.6	7.0	6.6	6.2	6.0	7.8	6.6	6.2
1+	9.6	9.3	10.9	10.9	10.9	12.1	9.5	10.0	9.3	9.3	9.3	9.3	9.5	10.0	9.3	9.3
2+	12.0	10.9	11.7	14.0	14.0	14.0	11.9	12.3	11.2	10.9	10.9	10.9	10.9	12.8	10.9	10.9
3+	14.5	12.4	14.2	15.7	15.7	16.2	14.4	14.2	12.8	12.4	12.4	12.4	12.4	14.2	12.4	12.4
4+	16.4	13.9	16.0	17.2	17.2	18.2	16.3	16.6	15.1	13.9	13.9	13.9	13.9	5.1	13.9	13.9
5+	17.4	15.5	17.8	18.5	18.5	19.7	17.3	18.5	16.8	16.0	16.0	15.5	15.5	16.8	15.5	15.5
6+	19.4	17.1	19.7	19.7	19.7	21.7	19.3	19.7	18.6	17.7	17.7	17.1	17.1	18.6	17.1	17.1
7+	21.4	18.7	23.3	21.6	21.6	24.0	21.3	21.6	21.6	20.4	21.6	18.7	18.7	20.4	18.7	18.7
8+	23.5	20.2	24.9	23.8	23.8	26.7	23.4	23.8	23.8	22.3	22.3	20.2	20.2	22.3	20.2	20.2
9+	25.2	22.8	27.5	27.5	26.2	28.7	25.1	26.2	24.3	24.3	24.3	21.8	22.8	24.3	21.8	21.8
10+	27.8	24.9	30.6	29.0	26.7	30.6	27.7	29.0	26.7	24.9	24.9	23.7	23.7	26.7	23.7	23.7
11+	30.8	27.5	34.3	32.4	29.7	34.3	30.6	29.7	29.7	27.5	27.5	26.1	26.1	29.7	26.1	26.1
12+	32.2	29.3	36.5	36.5	33.4	36.5	32.1	33.4	30.9	33.4	33.4	29.3	29.3	30.9	29.3	29.3
13+	34.8	33.4	41.4	38.0	38.0	41.4	34.6	35.2	35.2	35.2	35.2	33.4	33.4	35.2	33.4	33.4
14+	39.8	38.4	43.3	43.3	43.3	45.8	39.7	40.3	38.4	40.3	40.3	38.4	38.4	43.3	38.4	38.4
15+	44.9	43.4	48.5	48.5	48.5	49.2	44.7	45.4	43.4	45.4	45.4	43.4	43.4	45.4	43.4	43.4
16+	49.4	47.8	49.9	53.1	53.1	53.1	49.2	52.1	49.9	53.1	53.1	47.8	47.8	49.9	47.8	47.8
17+	53.1	51.0	56.3	56.3	56.3	56.3	52.8	60.3	51.0	56.3	56.3	51.0	51.0	53.1	51.0	51.0
Adult	57.5	55.6	64.6	58.2	60.2	62.9	57.4	60.5	56.5	58.2	59.1	53.1	53.9	55.4	51.1	55.7

Males: height cm

Age yrs	1	2	3	4	5	6	7	8	9	10	11	12	13	14	15	16
0+	63.5	62.8	64.4	66.4	66.4	67.8	63.4	69.2	66.4	66.4	66.4	62.8	64.4	65.6	65.6	62.8
1+	77.0	76.7	78.5	80.8	80.8	84.0	77.0	78.5	77.4	76.7	76.7	76.7	76.7	80.8	76.7	76.7
2+	87.5	83.8	85.9	92.3	92.3	93.4	87.5	82.5	83.8	83.8	83.8	83.8	83.8	87.5	83.8	83.8
3+	93.0	91.5	95.7	97.0	97.0	101.2	92.9	93.9	93.9	91.5	91.5	91.5	92.4	98.1	91.5	91.5
4+	99.0	98.2	100.9	102.8	102.8	108.9	98.9	100.9	104.2	98.2	98.2	98.2	98.2	102.8	98.2	98.2
5+	104.7	104.2	107.0	107.0	107.0	115.6	104.6	109.1	109.1	104.2	104.2	104.2	104.2	109.1	104.2	104.2
6+	110.1	112.6	114.8	112.6	112.6	121.6	109.1	113.3	116.3	112.6	112.6	109.6	109.6	112.6	109.6	109.6
7+	115.0	114.5	121.6	115.8	115.8	127.1	115.0	121.6	120.0	120.0	121.6	114.5	114.5	120.0	114.5	114.5
8+	119.8	119.2	124.9	120.5	120.5	128.2	119.7	126.7	126.7	124.9	126.7	119.2	119.2	124.9	119.2	119.2
9+	124.8	123.7	131.7	127.3	125.1	133.3	124.6	131.7	129.9	129.9	129.9	123.7	123.7	129.9	123.7	123.7
10+	129.0	128.3	137.0	129.8	132.1	137.0	128.9	137.0	135.0	135.0	135.0	128.3	128.3	132.1	128.3	128.3
11+	133.7	133.0	142.7	134.7	134.1	142.7	133.6	140.4	137.3	140.4	140.4	133.0	133.0	137.3	133.0	133.0
12+	138.9	138.1	146.4	140.0	142.9	146.4	138.8	142.9	142.9	146.4	142.9	138.1	138.1	142.9	138.1	138.1
13+	144.7	143.8	152.7	143.8	148.9	152.7	144.6	148.9	148.9	148.9	148.9	143.8	143.8	148.9	143.8	143.8
14+	151.9	150.4	155.4	152.4	155.4	164.1	151.7	152.4	155.4	155.4	155.4	150.4	150.4	152.4	150.4	150.4
15+	158.4	157.1	158.9	158.9	161.7	167.5	158.3	158.9	165.0	158.9	158.9	157.1	157.1	161.7	157.1	157.1
16+	163.9	162.3	162.3	163.9	166.4	169.4	163.7	166.0	169.4	163.9	163.9	162.3	162.3	163.9	162.3	162.3
17+	165.0	164.5	164.5	166.0	168.4	171.2	165.8	167.4	171.2	166.0	166.0	164.5	164.5	168.4	163.8	164.5
Adult	167.0	165.5	170.6	170.6	169.4	172.7	166.8	169.8	172.5	167.8	167.8	165.0	165.0	169.5	164.0	167.8

Females: weight kg

Age yrs	1	2	3	4	5	6	7	8	9	10	11	12	13	14	15	16
0+	5.9	6.1	6.8	7.0	6.5	6.8	5.9	7.0	6.1	6.5	6.1	6.1	5.6	6.5	6.1	6.1
1+	9.0	8.6	9.8	10.2	8.6	9.8	8.9	9.8	8.9	8.6	8.6	8.6	8.6	8.9	8.6	8.6
2+	11.2	10.5	11.9	11.9	10.5	12.3	11.1	11.9	10.8	10.5	10.5	10.5	10.5	11.3	10.5	10.5
3+	13.1	12.1	14.2	14.2	12.1	14.7	13.1	14.2	13.0	12.1	12.1	12.1	12.1	12.5	12.1	12.1
4+	14.6	13.4	15.9	15.9	13.8	16.8	14.6	15.3	15.3	13.8	13.8	13.4	13.4	13.8	13.4	13.4
5+	16.1	15.1	18.0	18.0	15.9	19.3	16.0	16.8	17.5	15.9	15.9	14.6	14.6	16.8	14.6	14.6
6+	17.6	16.5	19.3	19.3	18.5	21.5	17.6	19.3	19.3	18.5	18.5	15.9	15.9	19.3	15.9	15.9
7+	19.3	19.3	21.6	21.6	20.7	23.3	19.3	21.6	20.7	19.3	20.7	17.4	17.4	20.7	17.4	17.4
8+	21.6	20.2	24.5	24.5	23.3	26.6	21.6	23.3	23.3	23.3	23.3	19.2	19.2	23.3	19.2	19.2
9+	24.3	22.5	26.4	27.9	24.3	29.2	24.3	26.4	26.4	26.4	24.3	21.3	21.3	24.3	21.3	21.3
10+	25.2	25.2	29.8	31.7	27.3	33.3	25.2	27.3	29.8	27.3	27.3	23.8	23.8	27.3	23.8	23.8
11+	28.0	28.3	33.6	35.7	30.7	35.7	28.1	30.7	33.6	30.7	28.3	26.7	26.7	30.7	26.7	26.7
12+	30.0	31.7	37.6	40.0	37.6	40.0	30.0	34.4	34.4	34.4	34.4	30.0	30.0	34.4	30.0	30.0
13+	33.5	33.5	41.6	44.1	41.6	44.1	33.5	38.2	41.6	38.2	38.2	33.5	33.5	38.2	33.5	35.3
14+	36.8	38.8	45.3	47.8	45.3	47.8	36.8	45.3	45.3	41.7	41.7	36.8	38.8	41.7	36.8	36.8
15+	40.0	41.6	48.1	50.7	48.1	52.9	40.0	48.1	48.1	48.1	48.1	39.7	41.6	44.6	39.7	39.7
16+	41.9	43.5	49.8	52.3	49.8	54.4	41.9	52.3	49.8	52.3	49.8	41.6	43.5	46.4	41.6	41.6
17+	44.9	44.4	50.4	52.8	52.8	54.8	44.8	52.8	50.4	52.8	52.8	42.7	44.4	47.2	42.7	42.7
Adult	45.4	44.7	52.0	53.3	53.0	55.2	45.4	53.5	50.5	53.2	52.8	42.9	44.7	48.0	42.9	44.4

Females: height cm

Age yrs	1	2	3	4	5	6	7	8	9	10	11	12	13	14	15	16
0+	62.0	61.0	61.0	65.9	65.3	67.3	61.8	67.3	61.0	65.3	65.3	61.0	61.6	65.9	63.7	62.5
1+	77.2	75.1	77.0	80.9	75.1	81.7	77.2	77.0	75.1	75.1	75.1	75.1	75.1	75.4	75.1	75.1
2+	86.7	82.9	86.5	88.6	83.7	92.7	86.7	85.0	83.7	83.7	83.7	82.9	82.9	86.5	82.9	82.9
3+	92.0	90.6	94.7	97.0	90.6	100.0	92.0	92.9	92.9	90.6	90.6	90.6	90.6	95.9	90.6	90.6
4+	98.9	97.2	99.7	104.0	98.1	107.3	98.8	99.7	102.9	98.1	98.1	97.2	97.2	102.9	97.2	97.2
5+	103.6	103.9	107.6	107.6	102.8	114.0	103.5	105.6	110.4	103.9	103.7	102.8	102.8	109.1	102.8	102.8
6+	108.8	109.1	113.3	113.3	107.9	118.9	108.7	113.3	116.3	111.0	110.0	107.9	107.9	111.0	107.9	107.9
7+	114.8	114.0	118.6	118.6	114.0	123.5	114.7	120.5	120.5	118.6	120.5	112.6	112.6	118.6	112.6	112.6
8+	121.7	119.0	124.0	124.0	118.5	129.3	121.6	126.0	126.0	126.0	124.0	117.5	117.5	124.0	117.5	117.5
9+	127.2	124.2	129.6	126.7	124.2	135.2	127.1	131.7	131.7	131.7	129.6	122.6	122.6	126.7	122.6	122.6
10+	130.5	130.1	135.7	132.6	130.1	139.7	130.5	135.7	137.9	137.9	135.7	128.5	128.5	132.6	128.5	128.5
11+	135.5	135.2	139.3	139.3	136.8	144.6	135.4	139.3	142.4	142.4	142.4	135.2	135.2	139.3	135.2	135.2
12+	141.9	141.9	143.5	143.5	145.9	148.9	141.5	145.9	148.9	145.9	145.9	141.9	141.9	143.5	141.9	141.9
13+	145.8	146.5	148.1	150.5	150.5	153.4	146.5	153.4	153.4	150.5	150.5	146.5	146.5	148.1	146.5	146.5
14+	149.3	150.2	152.6	152.6	152.6	157.7	149.2	155.6	157.7	152.6	152.6	148.6	148.6	152.6	148.6	148.6
15+	152.0	151.0	153.5	156.5	153.5	158.6	151.9	156.6	158.2	156.5	156.5	149.5	149.5	153.5	149.5	149.5
16+	153.2	151.8	157.2	157.2	154.2	159.3	153.1	157.3	158.2	157.2	154.7	150.4	150.4	154.4	150.4	151.8
17+	154.7	152.0	158.2	158.2	155.5	160.2	154.6	158.2	158.6	158.2	155.5	151.8	151.8	155.5	151.8	152.0
Adult	155.3	152.5	158.7	158.7	156.1	160.6	155.1	158.7	158.7	158.7	156.1	152.5	152.5	156.1	152.5	156.1

Males: weight kg

Age yrs	17	18	19	20	21	22	23	24	25	26	27	28	29	30	31	32
0+	7.6	7.8	6.6	7.2	6.2	6.2	8.0	7.3	8.1	8.1	7.6	8.1	8.1	7.6	8.7	8.4
1+	10.5	11.2	9.8	10.2	9.3	9.3	12.1	10.5	11.5	11.5	10.5	11.5	11.5	10.5	12.1	11.8
2+	12.8	13.5	11.5	12.9	10.9	10.9	14.4	12.8	14.0	13.5	12.8	13.5	14.0	11.7	14.4	14.0
3+	14.2	16.2	12.8	14.6	12.4	12.4	16.2	14.8	16.2	15.7	14.8	15.7	15.7	13.4	16.7	15.7
4+	16.6	17.2	14.4	16.6	13.9	13.9	18.2	17.7	18.2	17.7	17.2	18.2	17.7	15.1	18.2	17.7
5+	17.8	18.5	16.1	17.7	15.5	15.5	20.3	19.7	20.3	19.7	19.1	19.7	19.7	17.8	20.3	19.7
6+	20.4	19.7	17.7	19.2	17.1	17.1	22.6	21.7	22.6	21.7	21.1	21.7	21.7	19.7	21.7	21.7
7+	22.5	21.6	19.4	21.0	18.7	18.7	25.1	25.1	25.1	24.0	23.3	24.0	23.3	21.6	25.1	25.1
8+	24.9	23.8	21.1	28.8	20.2	20.2	28.0	28.0	28.0	26.7	25.8	26.7	25.8	23.8	25.8	28.0
9+	27.5	26.2	22.9	26.2	21.8	21.8	29.7	31.4	31.4	29.7	28.7	29.7	28.7	26.2	29.7	29.7
10+	30.6	29.0	25.0	27.8	23.7	23.7	33.3	35.3	33.3	33.3	32.0	33.3	30.6	29.0	33.3	33.3
11+	34.3	32.4	26.7	29.7	26.1	26.1	37.5	39.8	37.5	35.9	35.9	37.5	34.3	34.3	35.9	35.9
12+	38.6	36.5	30.0	33.4	29.3	29.3	40.5	42.3	42.3	40.5	38.6	40.5	38.6	33.4	40.5	40.5
13+	43.8	38.0	34.2	38.0	33.4	33.4	45.9	47.8	47.8	45.9	43.8	45.9	43.8	38.0	43.8	45.9
14+	49.5	43.3	41.6	43.3	38.4	38.4	53.8	51.7	53.8	51.7	49.5	51.7	49.5	43.3	51.7	51.7
15+	52.3	48.5	45.4	47.0	43.4	43.4	59.5	55.0	59.5	57.3	57.3	59.5	55.0	48.5	57.3	57.3
16+	57.0	53.1	49.9	53.1	47.8	47.8	62.2	57.0	64.4	62.2	59.8	62.2	59.8	53.1	62.2	62.2
17+	60.3	56.3	51.9	56.3	51.0	51.0	65.5	60.3	67.8	65.5	63.1	65.5	63.1	56.3	65.5	65.5
Adult	62.5	58.0	55.6	62.1	53.9	58.8	78.5	69.2	71.2	76.3	75.0	71.2	71.2	61.9	77.2	72.5

Males: height cm

Age yrs	17	18	19	20	21	22	23	24	25	26	27	28	29	30	31	32
0+	67.1	65.6	66.0	65.4	62.8	62.8	70.1	67.8	70.1	69.2	67.8	67.8	69.2	67.1	70.1	69.2
1+	78.5	81.6	78.5	81.4	76.7	76.7	83.2	78.5	82.4	81.6	78.5	83.2	81.6	76.7	83.2	82.4
2+	87.5	85.9	85.9	90.5	83.8	83.8	93.4	88.6	92.3	92.3	89.5	92.3	92.3	83.8	93.4	92.3
3+	93.9	98.1	93.3	99.1	91.5	91.5	101.2	98.1	101.2	101.2	98.1	101.2	101.2	91.5	101.2	101.2
4+	100.9	102.8	99.4	106.6	98.2	98.2	110.3	105.4	108.9	105.4	104.2	108.9	105.4	98.2	108.9	108.9
5+	107.0	109.1	105.4	108.0	104.2	104.2	115.6	115.6	115.6	115.6	110.6	115.6	110.6	105.3	115.6	115.6
6+	114.8	112.6	110.9	113.7	109.6	109.6	121.6	121.6	121.6	121.6	116.3	121.6	116.3	109.6	121.6	121.6
7+	120.0	115.8	115.9	118.2	114.5	114.5	127.1	127.1	127.1	127.1	123.0	127.1	121.6	117.7	127.1	127.1
8+	124.9	122.5	120.6	123.5	119.2	119.2	132.4	132.4	132.4	132.4	126.7	132.4	126.7	122.5	132.4	132.4
9+	129.9	127.3	125.2	128.3	123.7	123.7	137.9	137.9	137.9	137.9	133.3	137.9	131.7	127.3	137.9	137.9
10+	135.0	132.1	129.9	134.5	128.3	128.3	143.7	143.7	143.7	143.7	138.7	143.7	137.0	132.1	143.7	143.7
11+	140.4	133.0	133.7	137.2	133.0	133.0	150.1	144.6	150.1	150.1	142.7	144.6	142.7	134.7	150.1	144.6
12+	146.4	142.9	138.9	144.6	138.1	138.1	157.2	151.0	151.0	151.0	148.9	151.0	148.9	142.9	151.0	151.0
13+	152.7	148.9	144.7	150.8	143.8	143.8	157.7	157.7	157.7	157.7	155.4	157.7	155.4	145.9	157.7	157.7
14+	159.1	152.4	151.3	157.2	150.4	150.4	170.6	164.1	164.1	170.6	164.1	164.1	161.8	150.4	164.1	164.1
15+	165.0	158.9	159.1	161.7	157.1	157.1	173.4	167.5	169.5	173.5	169.5	169.5	165.0	157.1	173.5	169.5
16+	166.0	163.9	163.0	165.9	162.3	162.3	175.5	170.6	173.4	175.4	173.3	173.4	169.4	162.3	175.4	173.4
17+	166.4	166.0	166.2	168.9	164.5	164.5	177.0	171.2	175.1	176.0	173.4	175.1	171.2	164.5	178.9	175.1
Adult	167.8	167.8	168.5	170.0	165.0	167.8	178.0	172.7	176.0	178.3	175.7	176.0	176.0	167.8	179.4	177.6

Females: weight kg

Age yrs	17	18	19	20	21	22	23	24	25	26	27	28	29	30	31	32
0+	7.0	7.0	6.1	6.3	6.1	6.1	8.0	7.2	7.4	7.7	7.2	7.2	7.2	7.0	8.0	7.7
1+	10.2	10.2	9.1	8.9	8.6	8.6	11.4	10.5	10.8	11.1	10.5	11.1	10.5	9.8	11.4	11.1
2+	11.9	12.3	11.1	10.9	10.5	10.5	13.9	13.5	13.0	13.0	12.3	13.0	13.0	11.3	13.9	13.5
3+	13.7	13.7	12.8	12.9	12.1	12.1	15.6	15.6	15.6	15.1	14.2	15.1	15.1	13.0	15.6	15.6
4+	15.3	15.3	14.4	14.5	13.4	13.4	17.5	17.5	17.5	16.8	16.4	17.5	16.8	15.3	18.2	18.2
5+	18.0	18.0	15.5	16.3	14.6	14.6	20.1	19.3	20.1	19.3	18.0	19.3	19.3	16.8	20.1	20.1
6+	20.0	19.3	17.4	18.5	16.5	15.9	21.5	21.5	22.6	21.5	20.6	21.5	21.5	18.5	21.5	22.6
7+	22.5	21.6	18.7	20.7	18.2	17.4	24.5	24.5	24.5	23.3	22.5	24.5	23.3	21.6	24.5	24.5
8+	24.5	23.3	20.8	22.5	20.2	19.2	26.6	26.6	28.1	26.6	25.6	26.6	25.6	23.3	26.6	26.6
9+	27.9	26.4	23.3	24.3	21.3	21.3	30.5	30.5	30.5	29.2	27.9	29.2	29.2	26.4	30.5	29.2
10+	29.8	29.8	25.3	28.5	23.8	23.8	33.3	33.3	34.7	31.7	31.7	33.3	33.3	27.3	33.3	33.3
11+	35.7	30.7	29.3	32.1	28.3	26.7	39.2	37.5	39.2	35.7	35.7	37.5	39.2	33.6	37.5	37.5
12+	40.0	37.6	31.9	37.6	31.7	30.0	43.8	42.0	43.8	42.0	40.0	43.8	46.7	37.6	42.0	42.0
13+	44.1	41.6	36.5	41.6	35.3	35.3	48.3	46.3	48.3	46.3	46.3	48.3	51.3	41.6	48.3	48.3
14+	47.8	41.7	40.0	45.3	38.8	36.8	52.1	50.0	52.1	52.1	50.0	52.1	52.1	45.3	52.1	52.1
15+	50.7	44.6	44.6	48.1	41.6	41.6	55.0	52.9	55.0	52.9	50.7	55.0	55.0	48.1	55.0	52.9
16+	52.3	46.4	46.4	49.8	43.5	43.5	56.4	54.4	56.4	54.4	52.3	56.4	56.4	52.3	56.4	56.4
17+	52.8	47.2	47.2	52.8	44.4	44.4	56.7	54.8	59.7	56.7	52.8	56.7	56.7	54.8	56.7	56.7
Adult	52.8	49.0	48.1	53.0	45.7	45.0	56.9	55.2	61.5	57.3	53.5	57.4	56.7	56.0	58.2	58.0

Females: height cm

Age yrs	17	18	19	20	21	22	23	24	25	26	27	28	29	30	31	32
0+	65.9	65.9	65.0	63.3	62.5	62.5	68.2	67.3	67.3	67.3	65.9	67.3	67.3	65.9	68.2	68.2
1+	77.0	75.1	77.3	76.6	75.1	75.1	81.7	78.3	80.1	82.5	78.3	80.9	78.3	75.9	81.7	80.1
2+	86.5	86.5	85.4	84.9	82.9	82.9	92.4	88.6	91.3	91.3	88.6	91.3	86.5	82.9	92.4	90.3
3+	92.9	92.9	92.4	94.5	90.6	90.6	100.0	97.9	98.9	98.9	97.0	98.9	95.9	90.6	100.0	98.9
4+	99.7	99.7	99.0	99.8	97.2	97.2	107.3	105.1	105.1	106.2	102.9	106.2	105.1	97.2	107.3	107.3
5+	107.6	107.6	104.0	107.3	102.8	102.8	114.0	112.7	112.7	111.6	109.1	112.7	110.4	103.9	114.0	114.0
6+	113.3	113.3	109.2	112.8	107.9	107.9	118.9	117.6	118.9	117.6	114.9	118.9	116.3	109.1	120.4	118.9
7+	120.5	118.6	114.1	119.6	112.6	112.6	124.9	123.5	124.9	123.5	120.5	123.5	122.0	116.1	126.5	124.9
8+	124.0	121.2	119.0	123.7	117.5	117.5	130.9	129.3	130.9	129.3	126.0	129.3	127.7	119.0	130.9	128.8
9+	129.6	126.7	124.4	130.5	122.6	122.6	136.9	135.2	135.2	135.2	131.7	133.5	131.7	126.7	136.9	133.5
10+	130.1	132.6	130.3	136.7	128.5	128.5	141.5	139.7	141.5	139.7	137.9	139.7	137.9	130.1	141.5	139.7
11+	142.4	135.2	135.9	143.5	135.2	135.2	148.2	146.4	148.2	146.4	144.6	146.4	142.4	136.8	146.4	146.4
12+	148.9	143.5	141.6	150.3	141.9	141.9	154.6	152.9	152.9	152.9	151.0	152.9	154.6	141.9	154.6	152.9
13+	152.4	150.5	146.5	155.0	146.5	146.5	159.0	155.5	159.0	159.0	157.3	157.3	157.3	148.1	159.0	157.3
14+	153.1	151.1	150.3	157.2	148.6	148.6	161.2	157.7	161.2	162.9	159.5	159.5	157.7	150.2	161.2	161.1
15+	153.5	152.6	151.3	158.2	149.5	149.5	162.1	158.6	162.1	163.8	162.1	162.1	160.4	153.5	163.8	161.2
16+	154.4	154.4	152.7	159.9	150.4	151.0	162.7	159.3	162.7	164.4	162.7	162.7	161.1	154.2	166.1	162.7
17+	155.5	155.5	155.1	160.0	151.8	152.8	163.4	160.2	163.4	165.0	163.4	163.4	162.2	155.5	166.6	163.8
Adult	156.1	156.1	156.1	160.5	152.5	156.1	163.7	160.6	163.7	165.7	164.2	163.7	163.7	156.1	166.8	165.2

Appendix 2.3 *Continued*

Males: weight kg

Age yrs	33	34	35	36	37	38	39	40	41	42	43	44	45	46	47	48
0+	8.4	7.8	7.8	8.1	7.0	7.6	7.6	7.0	7.6	7.6	7.0	7.3	6.0	7.0	7.0	7.1
1+	12.1	11.2	11.5	11.5	9.3	10.9	10.9	9.5	10.5	10.9	9.3	11.5	10.0	10.9	9.5	10.1
2+	14.4	13.5	13.5	13.5	11.2	13.2	12.8	10.9	12.3	12.8	11.2	13.5	12.8	12.8	10.9	12.6
3+	16.7	15.7	15.7	16.2	13.4	15.2	14.2	12.8	14.2	14.2	13.4	15.7	14.2	14.2	12.8	14.3
4+	18.8	18.2	17.7	17.7	15.1	16.6	16.0	15.1	16.0	16.0	15.1	17.7	16.0	16.0	15.1	16.5
5+	21.1	20.3	19.7	19.7	16.8	18.5	17.8	17.8	17.8	17.8	16.8	19.7	16.8	17.8	17.8	17.5
6+	23.5	21.7	21.7	22.6	20.4	21.1	19.7	19.7	19.7	19.7	20.4	21.7	18.6	19.7	19.7	19.0
7+	26.2	24.0	24.0	26.2	22.5	22.5	21.6	21.6	21.6	21.6	22.5	24.0	21.6	21.6	21.6	20.9
8+	28.0	26.7	25.8	29.4	24.9	24.9	24.9	24.9	24.9	23.8	24.9	26.7	23.8	24.9	24.9	23.8
9+	31.4	28.7	28.7	31.4	26.2	27.5	26.2	27.5	27.5	26.2	26.2	29.7	26.2	26.2	27.5	26.2
10+	35.3	33.3	32.0	35.3	29.0	30.6	30.6	30.6	29.0	29.0	29.0	32.0	26.7	30.6	30.6	28.1
11+	37.5	37.5	35.9	37.5	32.4	32.4	32.4	32.4	32.4	32.4	32.4	35.9	29.7	32.4	32.4	29.7
12+	42.3	42.3	40.5	42.3	33.4	33.4	36.5	36.5	36.5	36.5	33.4	38.6	33.4	36.5	36.5	33.4
13+	47.8	45.9	45.9	45.9	38.0	35.2	41.4	41.4	41.4	38.0	38.0	43.8	35.2	41.4	41.4	38.0
14+	53.8	51.7	51.7	51.7	40.3	40.3	46.9	43.3	46.9	43.3	40.3	46.9	38.4	46.9	43.3	43.3
15+	57.8	55.0	57.3	57.3	43.4	48.5	52.3	48.5	48.5	48.5	43.4	52.3	43.4	52.3	48.5	46.6
16+	62.2	59.8	59.8	59.8	49.9	53.1	53.1	53.1	53.1	53.1	49.9	53.1	49.9	53.1	53.1	53.1
17+	65.5	60.3	63.1	63.1	53.1	56.3	56.3	56.3	56.3	56.3	53.1	60.3	53.1	56.3	56.3	56.3
Adult	71.4	72.5	72.9	69.2	57.5	59.1	58.2	57.5	60.7	61.1	60.0	62.6	66.9	61.1	63.0	62.5

Males: height cm

Age yrs	33	34	35	36	37	38	39	40	41	42	43	44	45	46	47	48
0+	68.5	67.8	68.5	68.5	65.6	64.4	67.1	65.6	67.8	67.1	65.6	69.2	65.6	67.1	65.6	65.2
1+	82.4	80.8	82.4	81.6	76.7	79.8	80.8	76.7	79.8	80.8	76.7	84.0	77.4	80.8	76.7	78.3
2+	92.3	92.3	92.3	92.3	83.8	92.3	88.6	83.8	88.6	88.6	83.8	94.9	87.5	88.6	83.8	89.4
3+	101.2	101.2	98.1	101.2	91.5	97.0	97.0	91.5	97.0	97.0	91.5	101.2	95.7	97.0	91.5	97.0
4+	105.4	108.9	105.4	105.4	98.2	102.8	104.2	98.2	104.2	104.2	98.2	108.9	102.8	104.2	98.2	103.6
5+	115.6	115.6	111.9	110.6	104.2	109.1	109.1	104.2	110.6	110.6	104.2	115.6	107.0	109.1	104.2	108.3
6+	121.6	121.6	117.9	116.3	109.6	114.3	114.8	109.6	116.3	116.3	109.6	121.6	114.8	114.8	109.6	113.9
7+	127.1	127.1	123.0	123.0	114.5	120.0	120.0	114.5	121.6	121.6	114.5	127.1	120.0	120.0	114.5	119.3
8+	132.4	132.4	128.2	128.2	119.2	124.9	124.5	119.2	126.7	124.9	119.2	132.4	124.5	123.5	119.5	124.2
9+	137.9	137.9	137.9	133.3	123.7	129.9	129.9	123.7	131.7	131.7	123.7	137.9	127.3	129.9	123.7	129.1
10+	143.7	143.7	138.7	137.0	128.3	135.0	135.0	128.8	137.0	135.0	128.3	138.7	132.1	135.0	128.8	135.0
11+	144.6	144.6	144.6	140.4	133.0	140.4	137.3	134.7	140.4	140.4	133.0	142.7	137.3	137.3	134.7	137.9
12+	151.0	151.0	151.0	148.9	138.1	142.9	142.9	138.1	146.4	146.4	138.1	148.9	142.9	142.9	138.1	145.0
13+	157.7	155.4	157.7	152.7	143.8	145.9	148.9	143.8	152.7	152.7	143.8	155.4	145.9	148.9	143.8	151.2
14+	164.1	164.1	164.1	159.1	150.4	152.4	155.4	150.4	159.1	159.1	150.4	161.8	152.4	155.4	150.4	157.6
15+	169.5	169.5	169.5	165.0	157.1	158.4	161.7	157.1	165.0	161.7	157.1	167.5	157.1	161.7	157.1	161.5
16+	171.6	173.4	173.4	169.4	162.3	163.9	163.9	162.3	166.4	166.4	162.3	169.4	162.3	163.9	162.3	165.8
17+	173.3	175.1	174.1	171.2	164.1	166.0	168.4	164.5	168.4	168.4	164.5	171.2	167.5	168.4	164.5	169.0
Adult	173.6	177.6	174.4	173.4	165.5	167.8	170.6	165.5	170.6	170.6	165.5	172.7	174.4	170.6	165.5	170.1

Females: weight kg

Age yrs	33	34	35	36	37	38	39	40	41	42	43	44	45	46	47	48
0+	7.7	7.4	7.4	7.4	6.1	7.0	7.2	6.1	7.0	7.2	6.1	6.8	5.8	7.2	6.1	6.3
1+	11.1	11.1	11.1	10.5	8.9	10.5	10.5	8.6	9.8	10.5	8.9	10.2	9.8	10.5	8.6	9.0
2+	13.0	13.0	13.5	13.0	10.5	12.7	12.3	10.5	11.9	12.3	10.5	12.7	12.3	12.3	10.5	11.0
3+	16.2	15.6	15.1	15.6	13.0	14.7	14.7	12.1	13.7	14.2	13.0	15.1	13.7	14.7	12.1	13.1
4+	18.2	17.5	17.5	16.8	14.5	16.4	16.4	13.8	15.3	15.9	14.5	16.8	15.3	16.4	13.8	14.7
5+	20.2	19.3	19.3	19.3	16.8	18.0	17.5	15.9	17.5	17.5	16.8	19.3	16.8	17.5	15.9	16.4
6+	22.6	21.5	21.5	21.5	18.5	20.6	20.0	17.4	19.3	20.0	18.5	21.5	18.5	20.0	17.4	18.5
7+	25.7	24.5	23.3	24.5	19.3	22.5	21.6	19.3	21.6	21.6	19.3	23.3	20.7	21.6	19.3	20.7
8+	28.1	28.1	26.6	28.1	21.6	24.5	24.5	21.6	23.3	23.3	21.6	25.6	23.3	24.5	21.6	22.3
9+	30.5	30.5	29.2	32.4	24.3	26.4	27.9	22.5	26.4	26.4	24.3	29.2	26.4	27.9	22.5	24.3
10+	34.7	33.3	33.3	34.7	27.3	29.8	31.7	25.2	29.8	29.8	27.3	33.3	27.3	31.7	25.2	28.8
11+	39.2	37.5	37.5	39.2	30.7	35.7	35.7	28.3	33.6	33.6	30.7	37.5	30.7	35.7	28.3	32.4
12+	43.8	42.0	43.8	43.8	34.4	37.6	40.0	31.7	37.6	37.6	34.4	43.8	34.4	40.0	31.7	37.6
13+	48.3	46.3	48.3	46.3	38.2	41.6	41.6	35.3	41.6	41.6	38.2	46.3	41.6	41.6	35.3	41.6
14+	50.0	52.1	52.1	50.0	45.3	45.3	47.8	38.8	45.3	45.3	45.3	50.0	45.3	47.8	38.8	45.3
15+	52.9	55.0	55.0	52.9	48.1	48.1	48.1	44.6	48.1	48.1	48.1	52.9	48.1	48.1	44.6	48.1
16+	54.4	56.4	56.4	54.4	52.3	49.8	49.8	49.8	49.8	49.8	52.3	54.4	49.8	49.8	49.8	49.8
17+	54.8	56.7	56.7	54.8	52.8	50.4	50.4	50.4	50.4	50.4	52.8	54.8	52.8	50.4	50.4	52.8
Adult	56.4	58.0	56.7	55.6	53.0	50.9	51.0	50.9	50.4	50.5	50.5	53.0	55.1	53.2	51.0	53.0

Females: height cm

Age yrs	33	34	35	36	37	38	39	40	41	42	43	44	45	46	47	48
0+	65.9	65.3	65.3	67.3	61.0	63.7	65.3	61.0	66.6	65.3	61.0	68.2	61.6	65.3	63.7	63.6
1+	80.1	80.1	80.1	78.3	75.1	79.3	79.3	75.1	79.3	79.3	75.1	83.5	77.0	79.3	75.1	75.6
2+	88.6	90.3	88.6	89.5	82.9	87.6	86.5	82.9	87.6	87.6	82.9	93.9	88.6	86.5	82.9	83.4
3+	97.9	98.9	97.0	98.9	90.6	95.9	94.7	90.6	95.9	95.6	90.6	101.2	95.9	94.7	90.6	93.1
4+	106.2	105.1	104.0	105.1	97.2	101.5	105.1	97.2	101.5	104.0	97.2	107.3	102.9	104.0	97.2	100.1
5+	112.7	112.7	110.4	110.4	102.8	107.6	107.6	102.8	107.6	110.4	102.8	114.0	107.6	107.6	102.8	105.7
6+	118.9	117.6	116.3	116.3	107.9	113.3	113.3	107.9	114.9	116.3	107.9	116.3	113.3	113.3	107.9	113.6
7+	123.5	124.9	122.0	122.0	112.6	120.5	118.6	112.6	120.5	122.0	112.6	124.9	118.6	118.6	112.6	119.8
8+	127.7	129.3	127.7	127.7	117.5	124.0	124.0	117.5	124.0	126.0	117.5	129.3	124.0	124.0	117.5	121.8
9+	133.5	135.2	133.5	133.5	122.6	129.6	129.6	122.6	129.6	131.7	122.6	135.2	129.6	129.6	122.6	128.7
10+	139.7	139.7	139.7	139.7	128.5	132.6	135.7	128.5	135.7	137.9	128.5	139.7	135.7	135.7	128.5	134.8
11+	144.6	144.6	146.4	144.6	135.2	139.3	139.3	135.2	142.4	142.4	135.2	146.4	139.3	139.3	135.2	141.5
12+	152.9	152.9	152.9	148.9	141.9	143.5	145.9	141.9	145.9	148.9	141.9	154.6	145.9	145.9	141.9	147.7
13+	157.3	155.5	157.3	153.4	146.5	150.5	150.5	146.5	153.4	153.4	146.3	159.0	150.5	150.5	146.5	152.2
14+	159.5	159.5	158.7	156.5	148.6	152.6	152.6	148.6	155.6	155.6	148.6	159.5	155.6	152.6	148.6	154.5
15+	160.4	162.1	161.2	156.5	150.5	153.5	153.5	151.1	156.5	156.5	150.5	159.8	158.6	153.5	151.1	155.4
16+	161.1	162.7	161.8	157.2	152.4	154.4	157.2	152.0	157.3	157.2	152.4	160.0	161.1	157.2	152.0	156.1
17+	161.8	165.0	162.0	158.2	153.3	155.5	158.2	153.3	158.2	158.2	153.3	160.2	161.8	158.2	153.3	157.1
Adult	162.2	165.2	162.2	158.7	153.9	156.1	158.7	153.9	158.7	158.7	153.9	160.6	162.2	158.7	153.9	157.7

Males: weight kg

Age yrs	49	50	51	52	53	54	55	56	57	58	59	60	61	62
0+	7.6	7.6	7.8	7.3	7.6	7.6	7.8	7.6	7.8	8.1	7.7	7.8	6.6	7.8
1+	11.2	10.5	11.2	10.0	10.5	10.5	11.2	10.9	11.5	11.5	10.9	11.5	9.3	12.2
2+	13.2	12.3	14.0	10.9	12.8	11.7	14.0	13.2	13.5	14.0	13.6	13.5	10.9	14.6
3+	14.8	14.2	16.2	13.4	14.2	13.4	16.2	14.2	15.7	16.2	15.7	15.7	12.4	16.1
4+	17.2	16.0	18.2	14.4	16.6	15.1	18.2	16.6	17.7	17.7	17.2	17.7	13.9	18.0
5+	19.1	17.8	19.7	16.8	18.5	17.8	19.7	18.5	19.7	19.7	18.8	20.3	15.5	20.0
6+	21.1	18.6	21.1	18.6	21.1	19.7	21.7	20.4	21.7	22.6	20.4	22.6	17.1	22.0
7+	24.0	20.4	22.5	20.4	23.3	21.6	24.0	21.6	24.0	25.1	22.5	23.3	18.7	25.1
8+	25.8	22.3	24.9	22.3	25.8	23.8	28.0	24.9	26.7	28.0	24.4	28.0	20.2	28.0
9+	28.7	26.2	26.2	24.3	28.7	26.2	31.4	28.7	29.7	31.4	26.9	32.4	21.8	30.2
10+	30.6	26.7	29.0	26.7	30.6	29.0	35.3	30.6	33.3	33.3	27.9	35.3	23.7	35.1
11+	34.3	29.7	32.4	29.7	34.3	29.7	39.8	32.4	37.5	37.5	31.1	39.8	26.1	38.9
12+	38.6	33.4	36.5	33.4	38.6	33.4	45.0	38.6	42.3	42.3	33.4	45.0	29.3	43.2
13+	47.8	38.0	41.4	35.2	43.8	38.0	50.7	43.8	47.8	45.9	38.0	50.7	33.4	47.5
14+	49.5	40.3	46.9	40.3	46.9	43.3	56.9	46.9	53.8	51.7	43.3	56.9	38.4	52.4
15+	55.0	48.5	52.3	45.4	55.0	48.5	59.5	52.4	59.5	57.3	48.5	57.3	43.4	57.3
16+	57.0	53.1	57.0	49.9	57.0	53.1	64.4	57.1	64.4	62.2	53.1	59.8	47.8	62.1
17+	60.3	56.3	60.3	53.1	60.3	56.3	67.8	60.3	67.8	65.5	58.3	63.1	53.1	65.6
Adult	67.5	57.5	61.7	57.5	67.2	61.4	72.1	72.9	78.1	70.0	68.1	69.3	59.2	70.4

Males: height cm

Age yrs	49	50	51	52	53	54	55	56	57	58	59	60	61	62
0+	65.6	68.5	67.8	65.6	67.1	67.1	67.8	69.2	67.8	68.5	67.1	68.5	67.1	65.3
1+	76.7	77.4	80.8	76.7	80.8	76.3	80.8	80.8	82.4	80.8	78.4	82.4	76.7	84.6
2+	87.5	85.9	92.3	83.8	92.3	83.8	92.3	87.5	92.3	92.3	89.1	92.3	83.8	93.6
3+	95.7	93.9	98.1	91.5	98.1	91.5	98.1	95.7	101.3	101.2	96.4	98.1	91.5	100.0
4+	102.8	99.2	108.9	98.2	104.2	98.2	105.4	102.8	108.9	105.4	102.7	105.4	98.2	106.0
5+	109.1	104.2	115.6	104.2	109.1	105.3	111.9	107.0	115.6	111.9	108.1	111.9	104.2	111.9
6+	114.8	109.6	116.3	109.6	111.6	110.8	117.7	114.0	121.6	117.7	113.7	117.7	109.6	117.2
7+	121.6	115.8	120.0	117.7	114.5	117.7	127.1	117.7	127.1	127.1	117.9	123.0	114.5	123.9
8+	126.7	120.5	124.9	120.5	119.2	122.5	132.4	126.7	132.4	132.4	123.6	132.4	119.2	129.3
9+	131.7	125.1	129.9	123.7	129.9	127.3	137.9	129.9	137.9	137.9	127.5	137.9	123.7	133.7
10+	135.0	129.8	135.0	128.3	135.0	132.1	143.7	137.0	143.7	138.7	133.6	143.7	128.3	139.2
11+	140.4	134.7	137.3	134.7	140.4	134.7	150.1	140.4	150.1	144.6	135.7	144.6	133.0	145.1
12+	146.4	140.0	142.9	138.1	142.9	142.9	157.2	148.9	157.2	151.0	142.9	151.0	138.1	151.2
13+	152.7	145.9	148.9	143.8	148.9	145.9	164.3	152.7	164.3	155.4	148.9	157.7	143.8	156.8
14+	159.1	150.4	155.4	150.4	155.4	150.4	167.6	159.1	170.6	161.8	155.4	164.1	150.4	162.5
15+	165.0	158.9	161.7	157.1	161.7	157.1	169.5	165.0	175.5	167.5	160.3	169.5	157.1	167.7
16+	166.4	162.3	166.4	162.3	163.9	162.3	171.6	171.6	178.7	171.6	165.2	173.4	162.3	170.7
17+	168.0	164.5	168.4	164.5	166.0	164.5	173.3	173.3	180.1	173.3	168.4	174.1	163.5	174.1
Adult	170.6	165.5	170.6	165.5	170.6	167.8	175.1	174.4	180.4	174.4	171.9	174.4	164.0	175.0

Females: weight kg

Age yrs	49	50	51	52	53	54	55	56	57	58	59	60	61	62
0+	7.0	7.0	7.4	6.1	7.4	7.0	7.4	7.7	7.2	7.0	6.8	7.4	6.1	7.2
1+	10.5	9.8	11.1	8.6	11.1	9.8	11.1	11.1	10.8	10.5	9.4	11.1	8.6	11.4
2+	12.7	11.9	13.0	10.8	13.0	11.3	13.0	13.0	13.0	12.7	11.6	13.5	10.5	14.3
3+	14.7	13.0	15.1	12.5	14.7	13.0	15.1	14.7	15.1	14.7	13.2	15.1	12.1	16.0
4+	16.8	15.3	17.5	14.5	16.4	15.3	16.8	16.4	16.8	16.8	15.1	17.5	13.4	17.4
5+	18.0	16.8	19.3	15.9	18.6	16.8	19.3	17.5	18.6	19.3	17.3	19.3	14.6	19.5
6+	20.6	18.5	20.6	19.3	20.6	18.5	21.5	20.0	20.6	21.5	19.6	22.6	15.9	22.0
7+	23.3	20.7	22.5	20.7	22.5	21.6	23.3	22.5	23.3	24.5	21.6	24.5	17.4	25.0
8+	25.6	23.3	24.5	23.3	25.6	23.3	26.6	24.5	26.6	28.1	23.9	26.6	19.2	27.3
9+	30.5	26.4	26.4	24.3	27.9	26.4	32.4	27.9	30.5	30.5	25.4	30.5	21.3	31.0
10+	33.3	29.8	29.8	25.2	31.7	27.3	37.0	31.7	34.7	33.3	28.6	34.7	23.8	34.5
11+	35.7	33.6	33.6	28.3	33.6	33.6	41.9	35.7	39.2	37.5	32.2	39.2	26.7	38.6
12+	40.0	37.6	37.6	31.7	40.0	37.6	46.7	40.0	43.8	43.8	37.6	43.8	30.0	42.4
13+	46.3	41.6	44.1	38.2	44.1	41.6	51.3	44.1	48.3	48.3	41.6	48.3	33.5	47.1
14+	50.0	45.3	50.0	41.7	47.8	45.3	52.1	47.8	52.1	50.0	46.6	50.0	36.8	50.3
15+	52.9	48.1	52.9	48.1	52.9	48.1	55.0	50.7	55.0	52.9	50.5	52.9	39.7	52.2
16+	56.4	49.8	54.4	52.3	54.4	52.3	56.4	52.3	56.4	54.4	53.1	56.4	43.5	52.8
17+	56.7	50.4	56.7	54.8	54.8	54.8	56.7	54.8	56.7	54.8	56.4	56.7	47.2	54.0
Adult	57.6	52.8	56.7	55.7	56.1	56.0	56.8	55.9	56.7	55.3	56.8	56.7	47.3	61.4

Females: height cm

Age yrs	49	50	51	52	53	54	55	56	57	58	59	60	61	62
0+	63.7	67.3	65.3	64.5	65.9	65.9	65.3	68.2	65.9	67.3	65.6	65.3	65.9	63.7
1+	78.3	75.9	80.1	75.1	80.1	80.9	80.1	80.1	80.9	78.3	78.0	80.1	75.1	82.6
2+	86.5	85.0	92.4	82.9	92.4	82.9	92.4	87.6	89.5	87.6	84.4	88.6	82.9	92.2
3+	94.7	92.9	97.9	90.6	97.9	90.6	97.9	94.7	97.9	97.0	91.8	97.0	90.6	98.6
4+	102.9	99.7	106.2	97.2	104.0	97.2	106.2	102.9	105.1	104.0	99.8	104.0	97.2	105.1
5+	109.1	103.9	110.4	102.8	109.1	103.9	110.4	105.6	111.6	110.4	106.0	110.4	102.8	111.2
6+	113.3	111.0	114.9	109.1	113.1	109.1	117.6	114.9	117.6	116.3	111.4	116.3	107.9	116.1
7+	120.5	116.1	120.5	116.1	120.5	116.1	122.0	116.1	123.5	122.0	117.3	123.5	112.6	123.1
8+	124.0	121.2	124.0	121.2	124.0	119.0	127.7	126.0	129.3	127.7	122.3	129.3	117.5	127.6
9+	131.7	126.7	129.6	124.2	129.6	126.7	133.5	129.6	135.2	133.5	126.9	135.2	122.6	133.3
10+	135.7	132.6	137.9	128.5	135.7	130.1	141.5	137.9	141.5	137.9	132.9	141.5	128.5	139.5
11+	142.4	136.8	139.3	135.2	139.3	136.8	149.9	142.4	148.2	144.6	138.1	148.2	135.2	145.5
12+	148.9	143.5	145.9	141.9	143.5	141.9	154.6	145.9	154.6	151.0	145.9	154.6	141.9	151.4
13+	152.4	148.1	150.5	146.5	150.5	148.1	155.3	153.4	159.0	155.5	150.5	157.3	146.5	157.4
14+	153.6	149.2	155.6	148.6	155.6	150.2	156.5	157.7	161.2	157.7	152.6	159.5	148.6	159.7
15+	155.5	150.5	156.5	150.5	156.5	151.5	157.3	158.6	162.1	160.4	154.6	160.1	149.5	161.3
16+	157.2	151.4	157.2	152.0	157.2	153.2	157.9	161.1	162.7	161.1	156.8	161.7	150.4	162.0
17+	158.2	152.5	158.2	153.3	158.2	154.5	158.2	160.2	163.4	161.8	158.7	161.8	151.8	162.3
Adult	158.7	153.9	158.7	153.9	158.7	156.1	158.7	162.2	163.7	162.2	159.2	162.2	152.5	162.4

Appendix 3.1 NCHS reference weight (kg) for height values (0–9+ years)

1. MALES

Sex	Height in cm	3rd centile NCHS	5th centile NCHS	10th centile NCHS	20th centile NCHS	30th centile NCHS	40th centile NCHS	50th centile NCHS	60th centile NCHS	70th centile NCHS	80th centile NCHS	90th centile NCHS	95th centile NCHS
Male	56	3.2	3.4	3.7	4.0	4.3	4.5	4.7	5.0	5.3	5.7	6.2	6.7
Male	58	3.8	4.0	4.3	4.7	4.9	5.2	5.4	5.7	6.0	6.4	6.9	7.3
Male	60	4.4	4.6	4.9	5.3	5.6	5.8	6.0	6.3	6.6	7.0	7.6	8.0
Male	62	4.9	5.1	5.5	5.9	6.2	6.4	6.6	6.9	7.3	7.6	8.2	8.6
Male	64	5.4	5.7	6.0	6.4	6.7	7.0	7.2	7.5	7.8	8.2	8.8	9.2
Male	66	5.9	6.2	6.5	6.9	7.2	7.5	7.7	8.1	8.4	8.8	9.3	9.8
Male	68	6.4	6.7	7.0	7.4	7.8	8.0	8.3	8.6	8.9	9.3	9.9	10.4
Male	70	6.9	7.1	7.5	7.9	8.2	8.5	8.8	9.1	9.4	9.8	10.4	10.9
Male	72	7.4	7.6	8.0	8.4	8.7	9.0	9.2	9.6	9.9	10.3	10.9	11.4
Male	74	7.8	8.0	8.4	8.8	9.2	9.4	9.7	10.0	10.4	10.8	11.4	11.9
Male	76	8.2	8.5	8.8	9.3	9.6	9.9	10.1	10.5	10.9	11.3	11.9	12.4
Male	78	8.7	8.9	9.3	9.7	10.0	10.3	10.6	10.9	11.3	11.8	12.4	12.9
Male	80	9.1	9.3	9.7	10.1	10.5	10.8	11.0	11.4	11.8	12.2	12.8	13.4
Male	82	9.5	9.7	10.1	10.6	10.9	11.2	11.5	11.8	12.2	12.7	13.3	13.8
Male	84	9.9	10.1	10.5	11.0	11.3	11.6	11.9	12.3	12.7	13.1	13.8	14.3
Male	86	10.3	10.5	10.9	11.4	11.8	12.1	12.3	12.7	13.1	13.6	14.2	14.7
Male	88	10.7	10.9	11.3	11.8	12.2	12.5	12.8	13.2	13.6	14.0	14.7	15.2
Male	90	11.1	11.3	11.8	12.3	12.6	13.0	13.3	13.6	14.0	14.5	15.1	15.7
Male	92	11.5	11.8	12.2	12.7	13.1	13.4	13.7	14.1	14.5	15.0	15.6	16.1
Male	94	11.9	12.2	12.6	13.2	13.6	13.9	14.2	14.6	15.0	15.5	16.1	16.6
Male	96	12.3	12.6	13.1	13.6	14.0	14.4	14.7	15.1	15.5	16.0	16.6	17.1
Male	98	12.7	13.0	13.5	14.1	14.5	14.9	15.2	15.6	16.0	16.5	17.1	17.7
Male	100	13.2	13.5	14.0	14.6	15.0	15.4	15.7	16.1	16.5	17.0	17.7	18.2
Male	102	13.6	14.0	14.5	15.1	15.5	15.9	16.3	16.7	17.1	17.6	18.3	18.8
Male	104	14.1	14.4	15.0	15.6	16.1	16.5	16.9	17.3	17.7	18.2	18.9	19.5
Male	106	14.6	14.9	15.5	16.2	16.6	17.0	17.4	17.8	18.3	18.8	19.5	20.1
Male	108	15.1	15.5	16.0	16.7	17.2	17.6	18.0	18.5	18.9	19.5	20.2	20.8
Male	110	15.6	16.0	16.6	17.3	17.8	18.3	18.7	19.1	19.6	20.2	20.9	21.6
Male	112	16.2	16.6	17.2	17.9	18.4	18.9	19.3	19.8	20.3	20.9	21.7	22.4
Male	114	16.8	17.2	17.8	18.6	19.1	19.6	20.0	20.5	21.0	21.7	22.5	23.3
Male	116	17.4	17.8	18.5	19.2	19.8	20.2	20.7	21.2	21.8	22.5	23.4	24.2
Male	118	18.1	18.5	19.1	19.9	20.5	21.0	21.4	22.0	22.6	23.3	24.3	25.2
Male	120	18.8	19.2	19.9	20.7	21.2	21.7	22.2	22.8	23.5	24.2	25.3	26.2
Male	122	19.5	19.9	20.6	21.4	22.0	22.5	23.0	23.7	24.4	25.2	26.4	27.3
Male	124	20.2	20.7	21.4	22.2	22.8	23.4	23.9	24.6	25.3	26.2	27.5	28.5
Male	126	21.0	21.5	22.2	23.1	23.7	24.3	24.8	25.5	26.4	27.3	28.7	29.8
Male	128	21.8	22.3	23.0	24.0	24.6	25.2	25.7	26.6	27.5	28.5	30.0	31.2
Male	130	22.6	23.1	23.9	24.9	25.6	26.2	26.8	27.7	28.6	29.8	31.3	32.6
Male	132	23.4	24.0	24.8	25.9	26.6	27.3	27.8	28.8	29.9	31.1	32.8	34.2
Male	134	24.3	24.9	25.8	26.9	27.7	28.4	29.0	30.1	31.2	32.5	34.3	35.8
Male	136	25.2	25.8	26.8	28.0	28.8	29.6	30.2	31.4	32.6	34.0	36.0	37.6
Male	138	26.0	26.7	27.8	29.1	30.0	30.8	31.6	32.8	34.1	35.6	37.7	39.5
Male	140	27.0	27.7	28.9	30.3	31.3	32.2	33.0	34.3	35.7	37.3	39.6	41.5
Male	142	27.9	28.7	30.0	31.5	32.6	33.6	34.5	35.9	37.4	39.1	41.6	43.6
Male	144	28.8	29.7	31.1	32.8	34.0	35.1	36.1	37.6	39.2	41.1	43.7	45.8

2. FEMALES

Sex	Height in cm	3rd centile NCHS	5th centile NCHS	10th centile NCHS	20th centile NCHS	30th centile NCHS	40th centile NCHS	50th centile NCHS	60th centile NCHS	70th centile NCHS	80th centile NCHS	90th centile NCHS	95th centile NCHS
Female	56	3.3	3.5	3.7	4.1	4.3	4.5	4.7	5.0	5.3	5.7	6.2	6.7
Female	58	3.8	4.0	4.3	4.7	4.9	5.1	5.3	5.6	6.0	6.4	6.9	7.3
Female	60	4.4	4.6	4.9	5.3	5.5	5.8	6.0	6.3	6.6	7.0	7.5	8.0
Female	62	4.9	5.1	5.4	5.8	6.1	6.3	6.6	6.9	7.2	7.6	8.1	8.6
Female	64	5.4	5.6	5.9	6.3	6.6	6.9	7.1	7.4	7.8	8.2	8.7	9.1
Female	66	5.8	6.1	6.4	6.8	7.2	7.4	7.7	8.0	8.3	8.7	9.2	9.7
Female	68	6.3	6.5	6.9	7.3	7.6	7.9	8.2	8.5	8.8	9.2	9.7	10.2
Female	70	6.7	7.0	7.3	7.8	8.1	8.4	8.6	9.0	9.3	9.7	10.2	10.7
Female	72	7.2	7.4	7.8	8.2	8.6	8.8	9.1	9.4	9.7	10.1	10.7	11.1
Female	74	7.6	7.8	8.2	8.7	9.0	9.3	9.5	9.9	10.2	10.6	11.1	11.6
Female	76	8.0	8.3	8.6	9.1	9.4	9.7	10.0	10.3	10.6	11.0	11.6	12.1
Female	78	8.4	8.7	9.1	9.5	9.8	10.1	10.4	10.7	11.1	11.5	12.0	12.5
Female	80	8.8	9.1	9.5	9.9	10.2	10.5	10.8	11.1	11.5	11.9	12.5	12.9
Female	82	9.2	9.5	9.9	10.3	10.7	10.9	11.2	11.5	11.9	12.3	12.9	13.4
Female	84	9.6	9.9	10.3	10.7	11.1	11.4	11.6	12.0	12.3	12.8	13.4	13.8
Female	86	10.0	10.3	10.7	11.1	11.5	11.8	12.0	12.4	12.8	13.2	13.8	14.3
Female	88	10.4	10.7	11.1	11.6	11.9	12.2	12.5	12.8	13.2	13.7	14.3	14.8
Female	90	10.8	11.1	11.5	12.0	12.3	12.6	12.9	13.3	13.7	14.1	14.8	15.3
Female	92	11.2	11.5	11.9	12.4	12.8	13.1	13.4	13.8	14.2	14.6	15.3	15.8
Female	94	11.6	11.9	12.3	12.8	13.2	13.6	13.9	14.2	14.7	15.1	15.8	16.4
Female	96	12.0	12.3	12.7	13.3	13.7	14.0	14.3	14.7	15.2	15.7	16.3	16.9
Female	98	12.4	12.7	13.2	13.7	14.2	14.5	14.9	15.3	15.7	16.2	16.9	17.5
Female	100	12.8	13.1	13.6	14.2	14.7	15.0	15.4	15.8	16.2	16.8	17.5	18.1
Female	102	13.3	13.6	14.1	14.7	15.2	15.5	15.9	16.3	16.8	17.3	18.1	18.7
Female	104	13.7	14.1	14.6	15.2	15.7	16.1	16.5	16.9	17.4	17.9	18.7	19.4
Female	106	14.2	14.6	15.1	15.8	16.2	16.6	17.0	17.5	18.0	18.6	19.4	20.0
Female	108	14.7	15.1	15.6	16.3	16.8	17.2	17.6	18.1	18.6	19.2	20.0	20.7
Female	110	15.2	15.6	16.2	16.9	17.4	17.8	18.2	18.7	19.3	19.9	20.7	21.5
Female	112	15.8	16.2	16.8	17.5	18.0	18.5	18.9	19.4	19.9	20.6	21.5	22.2
Female	114	16.4	16.8	17.4	18.1	18.7	19.1	19.5	20.1	20.7	21.4	22.3	23.1
Female	116	17.0	17.4	18.0	18.8	19.3	19.8	20.3	20.8	21.4	22.2	23.2	24.0
Female	118	17.6	18.1	18.7	19.5	20.1	20.6	21.0	21.6	22.3	23.0	24.1	25.0
Female	120	18.3	18.7	19.4	20.2	20.8	21.3	21.8	22.5	23.2	24.0	25.1	26.1
Female	122	19.0	19.5	20.2	21.0	21.7	22.2	22.7	23.4	24.1	25.0	26.3	27.3
Female	124	19.7	20.2	21.0	21.9	22.5	23.1	23.6	24.4	25.2	26.2	27.5	28.6
Female	126	20.5	21.0	21.8	22.8	23.5	24.0	24.6	25.4	26.4	27.4	28.9	30.1
Female	128	21.3	21.9	22.7	23.7	24.5	25.1	25.7	26.6	27.6	28.8	30.4	31.8
Female	130	22.1	22.7	23.6	24.7	25.5	26.2	26.8	27.9	29.0	30.3	32.1	33.6
Female	132	23.0	23.6	24.6	25.8	26.6	27.4	28.0	29.2	30.5	31.9	33.9	35.6
Female	134	23.9	24.6	25.7	26.9	27.8	28.6	29.4	30.7	32.1	33.7	36.0	37.8
Female	136	24.9	25.6	26.7	28.1	29.1	30.0	30.8	32.2	33.8	35.6	38.2	40.3

Height (cm)	Weight (kg)	
	Male	Female
110	—	18.8
112	19.7	19.6
114	20.6	20.4
116	21.3	21.2
118	22.1	22.0
120	22.9	22.8
122	23.7	23.6
124	24.5	24.5
126	25.4	25.4
128	26.4	26.4
130	27.3	27.4
132	28.2	28.5
134	29.2	29.5
136	30.2	30.6
138	31.4	31.6
140	32.5	32.8
142	33.7	34.0
144	35.1	35.3
146	36.2	36.5
148	37.4	37.7
150	38.6	38.7
152	40.0	39.8
154	41.4	42.0
156	43.1	43.9
158	44.7	46.4
160	46.5	49.7
162	48.2	52.7
164	50.2	—
166	52.5	—
168	54.8	—
170	57.0	—
172	59.4	—
174	62.2	—

* For heights falling above or below the given range, use the predictive weight for height equations provided in Appendix 3.3.

Source: Baldwin, B. T. (1925). *Weight–height–Age standards in metric units for American-born children.* Am. J. Phys. Anth. **8**(1).

These equations provide an alternative approach to the estimation of weight for height and also a means of determining weight values when actual heights fall outside the ranges provided in Appendix 3.2.

The coefficients of regression for predicting the desirable body weight for height for the equation:

$$y = a + b(t^2) + c(t^3)*$$

Sex	Age (yrs)	Regression coefficients			
		(a) intercept	(b) for HT square	(c) for cubic HT	R²
Male	10+	19.9714	−0.00295	0.000026	0.9997
	11+	5.25937		0.0000098	0.9992
	12+	4.90330		0.0000100	0.9996
	13+	1.27530	0.00041	0.0000084	0.9997
	14+	− 5.72796	0.00118	0.0000055	0.9994
	15+	−35.04418	0.00471	−0.0000093	0.9989
	16+	−74.66398	0.00884	−0.0000250	0.9994
	17+	−61.18693	0.00784	−0.0000216	0.9986
Female	10+	22.1553	−0.00367	0.000030	0.9997
	11+	4.39673		0.0000105	0.9971
	12+	29.57905	−0.00378	0.0000284	0.9992
	13+	4.15742		0.0000109	0.9995
	14+	−76.26329	0.01095	0.0000376	0.9995
	15+	−66.58509	0.00959	−0.0000311	0.9907
	16+	−77.95227	0.01120	−0.0000380	0.9961
	17+	−61.72214	0.00936	−0.0000305	0.9975

* $y = a + b(t^2) + c(t^3)$

when y = resulting weight in kilograms.

t^2 and t^3 = actual height in centimetres squared and cubed, respectively.

Appendix 3.4 Fogarty reference weight for height values for adults*

Height without shoes (m)	Men Weight without clothes (kg)		Women Weight without clothes (kg)	
	Average	Weight range	Average	Weight range
1.45			46.0	42–53
1.48			46.5	42–54
1.50			47.0	43–55
1.52			48.5	44–57
1.54			49.5	44–58
1.56			50.4	45–58
1.58	55.8	51–64	51.3	46–59
1.60	57.6	52–65	52.6	48–61
1.62	58.6	53–66	54.0	49–62
1.64	59.6	54–67	55.4	50–64
1.66	60.6	55–69	56.8	51–65
1.68	61.7	56–71	58.1	52–66
1.70	63.5	58–73	60.0	53–67
1.72	65.0	59–74	61.3	55–69
1.74	66.5	60–75	62.6	56–70
1.76	68.0	62–77	64.0	58–72
1.78	69.4	64–79	65.3	59–74
1.80	71.0	65–80		
1.82	72.6	66–82		
1.84	74.2	67–84		
1.86	75.8	69–86		
1.88	77.6	71–88		
1.90	79.3	73–90		
1.92	81.0	75–93		

* Source: Bray, G. A., Ed. (1979). *Obesity in America*. Proceedings of the 2nd Fogarty International Center Conference on Obesity, Report No. 79, Washington, DC, Department of Health, Education and Welfare. Based on: 'Mortality among overweight men and women', Statistical Bulletin 41, New York, Metropolitan Life Insurance Co., 1960.

1. MALES

Sex	Age in yrs plus 6 months	3rd centile NCHS	5th centile NCHS	10th centile NCHS	20th centile NCHS	30th centile NCHS	40th centile NCHS	50th centile NCHS	60th centile NCHS	70th centile NCHS	80th centile NCHS	90th centile NCHS	95th centile NCHS
Male	0	6.0	6.2	6.6	7.0	7.3	7.6	7.8	8.1	8.4	8.7	9.1	9.4
Male	1	9.3	9.5	10.0	10.5	10.9	11.2	11.5	11.8	12.1	12.5	13.0	13.5
Male	2	10.9	11.2	11.7	12.3	12.8	13.2	13.5	14.0	14.4	15.0	15.7	16.4
Male	3	12.4	12.8	13.4	14.2	14.8	15.2	15.7	16.2	16.7	17.3	18.1	18.8
Male	4	13.9	14.4	15.1	16.0	16.6	17.2	17.7	18.2	18.8	19.5	20.5	21.3
Male	5	15.5	16.0	16.8	17.8	18.5	19.1	19.7	20.3	21.1	21.9	23.1	24.0
Male	6	17.1	17.7	18.6	19.7	20.4	21.1	21.7	22.6	23.5	24.5	25.9	27.1
Male	7	18.7	19.4	20.4	21.6	22.5	23.3	24.0	25.1	26.2	27.5	29.3	30.8
Male	8	20.2	21.0	22.3	23.8	24.9	25.8	26.7	28.0	29.4	31.1	33.4	35.3
Male	9	21.8	22.8	24.3	26.2	27.5	28.7	29.7	31.4	33.2	35.3	38.2	40.6
Male	10	23.7	24.9	26.7	29.0	30.6	32.0	33.3	35.3	37.5	40.0	43.6	46.5
Male	11	26.1	27.5	29.7	32.4	34.3	35.9	37.5	39.8	42.4	45.4	49.5	52.9
Male	12	29.3	30.9	33.4	36.5	38.6	40.5	42.3	45.0	47.8	51.2	55.9	59.8
Male	13	33.4	35.2	38.0	41.4	43.8	45.9	47.8	50.7	53.9	57.5	62.6	66.8
Male	14	38.4	40.3	43.3	46.9	49.5	51.7	53.8	56.9	60.2	64.0	69.4	73.8
Male	15	43.4	45.4	48.5	52.3	55.0	57.3	59.5	62.7	66.2	70.2	75.8	80.4
Male	16	47.8	49.9	53.1	57.0	59.8	62.2	64.4	67.7	71.3	75.5	81.3	86.1
Male	17	51.0	53.1	56.3	60.3	63.1	65.5	67.8	71.3	75.0	79.4	85.4	90.4
Male	18*	52.0	54.1	57.4	61.3	64.2	66.6	68.9	72.4	76.3	80.7	86.9	92.0

2. FEMALES

Sex	Age in yrs plus 6 months	3rd centile NCHS	5th centile NCHS	10th centile NCHS	20th centile NCHS	30th centile NCHS	40th centile NCHS	50th centile NCHS	60th centile NCHS	70th centile NCHS	80th centile NCHS	90th centile NCHS	95th centile NCHS
Female	0	5.6	5.8	6.1	6.5	6.8	7.0	7.2	7.4	7.7	8.0	8.4	8.7
Female	1	8.6	8.9	9.3	9.8	10.2	10.5	10.8	11.1	11.4	11.8	12.3	12.7
Female	2	10.5	10.8	11.3	11.9	12.3	12.7	13.0	13.5	13.9	14.5	15.2	15.8
Female	3	12.1	12.5	13.0	13.7	14.2	14.7	15.1	15.6	16.2	16.9	17.8	18.6
Female	4	13.4	13.8	14.5	15.3	15.9	16.4	16.8	17.5	18.2	19.0	20.1	21.0
Female	5	14.6	15.1	15.9	16.8	17.5	18.0	18.6	19.3	20.1	21.1	22.4	23.6
Female	6	15.9	16.5	17.4	18.5	19.3	20.0	20.6	21.5	22.6	23.7	25.4	26.7
Female	7	17.4	18.2	19.3	20.7	21.6	22.5	23.3	24.5	25.7	27.2	29.3	31.0
Female	8	19.2	20.2	21.6	23.3	24.5	25.6	26.6	28.1	29.8	31.7	34.4	36.6
Female	9	21.3	22.5	24.3	26.4	27.9	29.2	30.5	32.4	34.4	36.8	40.2	42.9
Female	10	23.8	25.2	27.3	29.8	31.7	33.3	34.7	37.0	39.5	42.3	46.3	49.6
Female	11	26.7	28.3	30.7	33.6	35.7	37.5	39.2	41.9	44.7	48.0	52.5	56.3
Female	12	30.0	31.7	34.4	37.6	40.0	42.0	43.8	46.7	49.8	53.4	58.4	62.5
Female	13	33.5	35.3	38.2	41.6	44.1	46.3	48.3	51.3	54.5	58.3	63.5	67.9
Female	14	36.8	38.8	41.7	45.3	47.8	50.0	52.1	55.2	58.5	62.4	67.8	72.2
Female	15	39.7	41.6	44.6	48.1	50.7	52.9	55.0	58.1	61.4	65.3	70.8	75.3
Female	16	41.6	43.5	46.4	49.8	52.3	54.4	56.4	59.6	62.9	66.9	72.3	76.8
Female	17	42.7	44.4	47.2	50.4	52.8	54.8	56.7	59.9	63.2	67.1	72.6	77.1
Female	18*	42.9	44.7	47.3	50.5	52.8	54.8	56.6	59.7	63.1	67.0	72.5	76.9

* Age is 18 years plus 0 months.

Appendix 4.1 Energy cost of physical activity classified by Physical Activity Ratio*

Note: activity descriptions are taken from the papers surveyed and there is some repetitiveness arising from this.

Physical Activity Ratio	Activities grouped in order of increasing ranges of PAR
1.2 (1.0–1.4)	Lying at rest Sitting at rest Standing at rest Sitting (unspecified activities) Laboratory work Sitting (reading/writing/calculating) Playing cards Lying reading Standing dissecting Laboratory work—sitting activities Resting in squatting position Sitting eating Piloting an aeroplane Sitting listening to radio
1.6 (1.4–1.8)	Washing, dressing, shaving, etc. Preparing vegetables Ironing (2.5 kg iron) Standing (unspecified activities) Knitting Sewing by hand Ironing General office work Washing dishes General laboratory work Sitting in the back of an army truck Sitting singing Standing singing Sitting playing with toys Walking on level at 2–3 km/h Playing the piano
2.1 (1.8–2.4)	Household chores (unspecified) Cooking activities (general) Operating foot driven sewing machine Standing drawing on wall Carpentry Playing darts Standing to attention Making tortillas—cooking Carpet weaving/weaving Walking on level 3–4 km/h, 20 kg load Walking on level 3–4 km/h, 20 kg load, altitude 2640 m Walking on level 3–4 km/h, 20 kg load, altitude 4560 m
2.8 (2.4–3.3)	Sitting activities (cooking, washing, etc.) Musketry (this uses the arms only) Cleaning/brushing boots/shoes, polishing army kit Playing pool Walking on level (3–4 km/h) Walking downstairs Archery Sweeping floor

* Schofield, E. C., Dallosso, H. M., and James, W. P. T. unpublished data.

Physical Activity Ratio	Activities grouped in order of increasing ranges of PAR
	Washing towels
	Laundry (washing, wringing, rinsing, hanging on the line)
	Building a jungle camp
	Taking a shower
	Getting dressed and undressed
	Step test (30 cm step at 12 steps/min)
	Army field craft
	Standing picking coffee/picking cotton
	Winnowing
	Standing cutting tobacco/harvesting sorghum/maize
	Walking downhill, speed unspecified, no load
	Walking on level 4–5 km/h, 20 kg load
	Walking on level 4–5 km/h, 20 kg load, altitude 2640 m
3.8 (3.3–4.4)	Walking on level (speed unspecified)
	Walking on level (5–6 km/h)
	Walking uphill (4–5 km/h)
	Cycling (on flat ground, 'own' speed)
	Walking on level (4–5 km/h)
	Weeding (moderate crop of weeds)
	Clearing land
	Cricket—bowling
	Competition drill
	Weapon training
	Gardening
	Golf
	Washing floor
	Playing table tennis
	Bed making
	Walking up and down stairs (rate unspecified)
	Marching at 5 km/h
	Walking (speed unspecified), 5–15 kg load
	Mopping the floor
	Calisthenics
	Walking (speed unspecified), 20 kg load (on head)
	Walking (speed unspecified), 30 kg load (on head)
	Building irrigation channels
	Harvesting rice
	Walking 3–4 km/h, 14 kg load (on head)
	Walking 4–5 km/h, 14 kg load (on head)
	Gymnastics
	Army patrol
	Gleaning
	Standing planting tobacco/taro/sweet potato/maize/root crops
	Raking gravel
	Sowing
	Grinding grain in millstone/grinding maize
	Fetching water from well/drawing water
	Walking downhill at own pace
	Harvesting root crops
	Walking on level 5–6 km/h, 20 kg load
	Walking on level 4–5 km/h, 20 kg load, altitude 4560 m
	Walking on level 5–6 km/h, 20 kg load, altitude 4560 m
5.1 (4.4–5.9)	Squad or Platoon Drill (incl. standing, saluting, marching, turning)
	Marching with 20 kg load—at ease
	Pounding rice
	Hoeing
	Marching on level at 5.5 km/h
	Potted sports (relay race, sprint, rope climbing, vaulting, rest periods)
	Running on level (less than 9 km/h)
	Misc. unspecified games (incl. volley ball, table tennis, push ball)
	Polishing furniture

Physical Activity Ratio	Activities grouped in order of increasing ranges of PAR
	Marching (speed unspecified) with full army pack
	Walking (speed unspecified) 15–25 kg load
	Walking (speed unspecified) 25–35 kg load
	Walking at 5–6 km/h pushing 50 kg cart
	Walking downhill (6–7 km/h) carrying 24 kg load
	Rebound running (jogging on a trampoline!)
	Cutting grass (with a scythe)
	Bush clearing
	Step test (30 cm step, 24 steps/min)
	Ploughing (unmechanized)
	Threshing rice
	Walking 3–4 km/h, 35 kg load (distrib.)
	Walking 3–4 km/h, 50 kg load (distrib.)
	Walking 3–4 km/h, 60 kg load (distrib.)
	Walking 3–4 km/h, 70 kg load (distrib.)
	Walking 5–6 km/h, 14 kg load (on head)
	Army assault course
	Playing cricket (nets)
	Digging trenches
	Dancing
	Swimming—stroke unspecified
	Road construction
	Digging holes/digging ground/shovelling mud
	Fire fighting (army)
	Walking uphill, speed unspecified, no load
	Chopping wood/cutting bamboo/splitting wooden posts/felling trees
	Walking on level 5–6 km/h, 20 kg load, altitude 2640 m
	Walking on level 6–7 km/h, 20 kg load, altitude 2640 m
	Walking on level 6–7 km/h, 20 kg load, altitude 4560 m
	Walking on level 7–8 km/h, 20 kg load, altitude 4560 m
6.7 (5.9–7.9)	Walking on level (6–7 km/h)
	Climbing stairs (rate unspecified)
	Parade drill
	Playing football
	Running (speed unspecified)
	Swimming (crawl stroke)
	Walking (speed unspecified) carrying 35–45 kg load
	Walking uphill (5–6 km/h) carrying 24 kg load
	Cross country walking—uphill and downhill
	Walking 4–5 km/h, 35 kg load (distrib.)
	Walking 4–5 km/h, 50 kg load (distrib.)
	Walking 4–5 km/h, 60 kg load (distrib.)
	Walking 4–5 km/h, 70 kg load (distrib.)
	Cutting sugar cane
	Paddling a 2-man canoe (competitively)
	Playing lacrosse
	Sailing
	Mowing the lawn
	Loading sacks of cotton/grain on to lorry
	Walking on level 7–8 km/h, 20 kg load, altitude 2640 m
9.0 (7.9–10.5)	Walking on level (7–8 km/h)
	Walking on level (6–7 km/h) carrying 20 kg load
	Walking on level (7–8 km/h) carrying 20 kg load
	Walking 6–7 km/h, 35 kg load (distrib.)
	Playing tennis (competitively)
	Boxing

Activity description	Cost kJ/min	Physical activity ratio	No. subjects
Aerobic dancing—low intensity ('equivalent to walking')	16.98	3.91	4
Aerobic dancing—medium intensity ('equivalent to jogging')	27.50	6.31	4
Aerobic dancing—high intensity ('equivalent to running')	35.94	8.21	4
Archery	21.91	4.35	4
Army obstacle course (water jump, scaling wall, rests, etc.)	25.76	4.95	6
Bed making	20.42	4.57	5
Bowls	18.40	3.75	1
Carrying 20–30 kg (jungle patrol)	16.62	3.42	7
Carrying 20–30 kg (jungle march)	25.52	5.09	2
Cleaning/brushing boots/shoes, polishing army kit	11.48	2.11	9
Cleaning windows	12.36	2.72	1
Cleaning stairs	14.01	3.21	20
Cleaning and drying	19.79	5.13	5
Climbing mountains at own pace at altitude of 6000 ft	41.67	7.98	2
Climbing mountains at own pace at altitude of 15 000 ft	36.58	6.89	3
Climbing mountains at own pace at altitude of 20 000 ft	35.58	6.64	3
Climbing stairs (72 steps/min)	20.11	5.19	8
Climbing stairs (92 steps/min)	27.31	6.95	3
Climbing stairs (rate unspecified)	35.76	6.79	22
Climbing stairs (rate unspecified) carrying 'light' load/basket of clothes	35.56	7.09	2
Climbing stairs (rate unspecified) carrying 11–16 kg load	41.11	6.83	4
Coal getting	29.97	5.65	10
Coal mining activities, average cost (excluding rest periods)	29.66	5.62	12
Cooking activities (general)	6.73	1.80	15
Cricket—batting	33.47	6.30	4
Cricket—bowling	33.47	6.35	4
Crocheting	4.79	1.17	1
Cycling (on flat ground, 'own' speed)	26.27	5.50	82
Dancing—Eightsome reel	32.22	5.20	3
Dancing—Foxtrot	22.99	3.57	2
Dancing—Waltz	26.77	4.22	2
Dressing doll	5.83	1.42	1
Driving (15 cwt truck)	14.22	2.89	3
Drill, Squad of Platoon (incl. standing, saluting, marching, turning)	15.15	3.14	8
Drill—church parade	17.08	3.38	9
Drill—parade	17.08	3.38	9
Drill—competition	25.89	5.08	11
Drill—competition practice	21.57	4.24	11
Drilling a shot hole (coal-mining)	29.69	5.15	2
Dusting	14.55	3.66	4
Field ridging	37.40	8.33	5
Finishing copper bands	14.34	3.78	2
Forging	13.07	3.22	4
Gardening	22.47	4.82	11
Gauging	15.59	4.18	4
Golf	20.90	4.40	1
Hewing	28.85	6.07	18
Hoeing	24.22	6.53	3
Hoisting shell (using pulley)	13.96	3.68	1
Household chores (unspecified)	13.68	3.65	1
Ironing (2.5 kg iron)	17.54	3.50	5
Ironing	5.97	1.46	1
Kneeling scrubbing (i.e. the floor)	9.54	2.68	3
Knitting	4.93	1.20	1

* Schofield, E. C., Dallosso, H. M., and James, W. P. T. unpublished data.

Activity description	Cost kJ/min	Physical activity ratio	No. subjects
Laboratory work	7.43	2.11	1
Laboratory work (general)	14.32	2.38	5
Labouring	21.10	5.39	5
Laundry (washing, wringing, rinsing, hanging on the line)	12.78	3.19	4
Leg exercises (lying on back lifting 9.5 kg load with legs)	9.01	1.68	12
General laboratory work	14.32	2.38	5
Loading	30.38	6.13	31
Lying at rest	4.96	1.05	229
Marching with 15 kg load	23.73	4.64	3
Marching with 20 kg load—at ease	23.96	5.08	5
Marching with 20 kg load—to attention	23.66	5.14	4
Marching on level at 5.5 km/h	24.01	4.70	13
Motor cycling on level ground	11.81	2.39	3
Musketry (this uses the arms only)	10.95	2.36	6
Office work (general)	5.74	1.18	3
Operating foot driven sewing machine	6.81	1.42	1
Operating electric sewing machine	5.07	1.06	1
Planting groundnuts	15.19	3.08	4
Playing pool	13.89	3.02	4
Playing squash	42.59	8.50	6
Playing tennis	29.84	5.83	7
Playing football	29.31	5.99	2
Playing table tennis	19.31	3.66	1
Playing cards	8.88	1.48	5
Pounding rice	21.54	5.59	4
Preparing vegetables	5.93	1.56	12
Pulling a rickshaw (180 kg load, this is a typical load)	42.53	10.40	21
Rifle cleaning (involves sitting and standing activities)	11.24	2.25	6
Running downhill (6% gradient) at 12 km/h (treadmill)	40.26	8.16	10
Running (speed unspecified)	39.83	7.61	4
Scrubbing floor	14.38	3.17	1
Setting the table	13.51	3.37	4
Sewing by hand	4.93	1.11	2
Sitting at rest	5.50	1.19	172
Sitting working (according to occupation)	6.44	1.44	47
Sitting activities (cooking, washing etc.)	8.69	2.13	11
Sitting typing	6.76	1.69	5
Sitting (unspecified activities)	6.56	1.33	47
Sitting activities (coal-mining)	6.63	1.39	19
Sitting (reading/writing/calculating)	8.18	1.36	5
Sitting typing—electric typewriter	4.88	1.35	3
Skiing (cross-country 25–35 kg army pack 4.5 km/h)	37.69	7.88	9
Sports assorted (relay race, sprint, rope climbing, vaulting, rest periods)	13.59	2.65	12
Standing at rest	5.96	1.26	156
Standing working (according to occupation)	7.46	1.71	51
Standing activities (washing rickshaws)	9.76	2.40	11
Standing (unspecified activities)	7.17	1.51	37
Standing activities (coal-mining)	8.27	1.72	19
Standing scrubbing (i.e. a table top)	8.73	2.45	3
Stamping	13.51	3.72	2
Step test (25 cm step at 15 steps/min)	20.17	4.17	50
Stone cutting	30.89	7.51	6
Stone splitting (sitting)	12.12	2.95	4
Stone splitting (standing)	15.35	3.72	4
Sweeping floor	7.01	1.71	1
Timbering	25.05	5.10	26
Tool setting	12.85	3.42	5
Trimming and stripping bark off a tree (using axe and spade)	39.89	7.99	12
Turning and finishing	11.51	3.02	8
Turning (heavy)	12.79	3.40	5

Activity description	Cost kJ/min	Physical activity ratio	No. subjects
Turning (light)	11.44	2.96	8
Walking on level (speed unspecified)	18.04	3.88	105
Walking on level (1–2 km/h)	9.47	2.23	20
Walking on level (3–4 km/h)	12.68	2.87	30
Walking on level (4–5 km/h)	17.87	3.77	83
Walking on level (5–6 km/h)	20.40	4.17	207
Walking on level (6–7 km/h)	27.40	5.47	63
Walking on level (7–8 km/h)	44.61	7.45	5
Walking on level (4–5 km/h) carrying 8 kg load	20.06	5.13	6
Walking (coal-mining)	25.33	5.31	19
Walking and carrying	16.32	3.98	3
Walking 6–7 km/h carrying 70 kg load (distrib.)	32.36	7.25	1
Walking uphill (4–5 km/h)	29.38	6.59	20
Walking at speed unspecified carrying 11–16 kg load	27.92	4.64	4
Walking uphill at own pace at altitude of 6000 ft	45.56	8.83	1
Walking uphill at own pace at altitude of 15 000 ft	40.74	7.67	3
Walking uphill at own pace at altitude of 20 000 ft	39.74	7.43	3
Walking downhill (8% gradient) at 1–2 km/h	28.76	1.85	10
Walking downhill (8% gradient) at 3–4 km/h	9.87	2.09	10
Walking downhill (8% gradient) at 4–5 km/h	11.83	2.51	10
Walking downhill (8% gradient) at 6–7 km/h	16.10	3.41	10
Walking downstairs (rate unspecified) carrying 'light' load/basket of clothes	16.74	3.36	2
Walking downstairs (rate unspecified) with 15 kg load	17.95	3.60	2
Walking downstairs (rate unspecified) with 30 kg load	22.99	4.60	2
Walking up and down stairs (97/min)	38.99	6.30	4
Walking up and down stairs (116/min)	41.35	6.87	5
Washing, dressing, shaving, etc.	13.47	2.69	64
Washing floor	6.18	1.51	1
Washing towels	7.71	1.88	1
Washing dishes	6.32	1.54	1
Watching football	8.94	1.84	5
Weapon training	9.25	1.82	12
Weeding (moderate crop of weeds)	23.38	5.03	3
Working on car assembly line (variety of unspecified tasks)	12.31	2.36	13

Activity: Task Description	Light PAR from 1985 Report 1.0–2.5	I.E.I. estimate	Task description	Moderate PAR from 1985 Report 2.6–3.9	I.E.I. estimate	Task description	Heavy PAR from 1985 Report 4.0+	I.E.I. estimate
MALES								
Household								
Lying	1.2	—	Strolling	2.5	2.27	Chopping firewood	4.1	3.08
Sitting quietly	1.2	—	Fishing with spear	2.6	2.34	Laying floor (LDC)	4.1	3.08
Standing quietly	1.4	—	Light cleaning	2.7	2.42	Walking uphill slowly	4.7	3.44
Cooking	1.8	1.61	Tying fence posts	2.7	2.42	Walking uphill— normal pace	5.7	4.04
Fishing with line	2.1	1.69	Walking slowly	2.8	2.49	Heavy reacreational— jogging, athletics	6.6	4.58
Fishing from canoe	2.2	1.71	Walking downhill slowly	2.8	2.49	Walking uphill fast	7.5	5.12
Playing cards	2.2	1.71	Weaving bamboo wall	2.9	2.57			
Washing clothes	2.2	1.71	Roofing house	2.9	2.57			
Making bows and arrows	2.7	1.84	Walking downhill— normal pace	3.1	2.74			
Light recreational— billiards, golf, cricket	2.2–4.4	1.71–2.26	Singing and dancing	3.2	2.79			
			Walking at normal pace	3.2	2.79			
			Nailing	3.3	2.87			
			Hunting flying fox	3.4	2.94			
			Hunting birds	3.4	2.94			
			Walking with 10 kg load	3.5	3.02			
			Hunting pigs	3.6	3.09			
			Walking downhill fast	3.6	3.09			
			Moderate cleaning	3.7	3.17			
			Moderate recreational— dancing, swimming, tennis	4.4–6.6	3.69–5.34			
Occupational								
Office work	1.3	—	Shoemaking	2.6	2.34	Putting coconuts in bags	4.0	3.02
Driving lorry	1.4	—	Kneading clay	2.7	2.42	Brick breaking	4.0	3.02
Flying helicopter	1.5	—	Painting and decorating	2.8	2.49	Sharpening posts	4.0	3.02
Sewing	1.5	1.54	Planting	2.9	2.57	Planting trees	4.1	3.08
Hovering in helicopter	1.6	1.56	Milking cows by hand	2.9	2.57	Cutting palm tree trunks	4.1	3.08
Sorting crops, kneeling	1.6	1.56	Making bricks, squatting	3.0	2.64	Splitting wood for posts	4.2	3.14
Moving around office	1.6	1.56	Electrical industry	3.1	2.72	Sawing and power sawing	4.2	3.14
Pre-flight check: pilot	1.8	1.61	Machine tool industry	3.1	2.72	Route marching (Army)	4.4	3.26
Laboratory work	2.0	1.66	Cutting bamboo	3.2	2.79	Shovelling mud	4.4	3.26
Weaving	2.1	1.69	Joinery	3.2	2.79	Collecting coconuts incl. climbing trees	4.6	3.38
Carving	2.1	1.69	Drill (Army)	3.2	2.79	Cutting grass with machete	4.7	3.44
Driving tractor	2.1	1.69	Bricklaying	3.3	2.87	Loading sacks— mechanized	4.7	3.44
Fishing with line	2.1	1.69	Paddling canoe	3.4	2.94	Cutting trees	4.8	3.50
Sorghum harvest— cutting ears	2.1	1.69	Jungle patrol (Army)	3.5	3.02	Pushing wheelbarrow	4.8	3.50
Cleaning kit (Army)	2.4	1.76	Uprooting timbers	3.5	3.02	Repairing fences	5.0	3.62
Tailoring	2.5	1.78	Carpentry	3.5	3.02	Digging holes for posts	5.0	3.62
			Chemical industry	3.5	3.02	Assault course (Army)	5.1	3.68
			Feeding animals	3.6	3.09	Labouring	5.2	3.74
			Making a fence	3.6	3.09	Collecting & spreading manure by hand	5.2	3.74
			Lifting grain sacks	3.7	3.17	Pulling cart without load	5.3	3.80
			Winnowing	3.9	3.32	Digging irrigation channels	5.5	3.92

* FAO/WHO/UNU (1985). *Energy and protein requirements. Report of a Joint Expert Consultation.* WHO Technical Report Series, No. 724, WHO, Geneva.

Appendix 5.1 *Continued*

Activity: Task Description	Light PAR from 1985 Report 1.0–2.5	I.E.I. estimate	Task description	Moderate PAR from 1985 Report 2.6–3.9	I.E.I. estimate	Task description	Heavy PAR from 1985 Report 4.0+	I.E.I. estimate
MALES (*continued*)								
Occupational (continued)						Digging earth to make mud	5.7	4.04
						Shovelling	5.7	4.04
						Jungle march (Army)	5.7	4.04
						Pulling cart with load	5.9	4.16
						Mining with pick	6.0	4.22
						Earth cutting	6.2	4.34
						Digging holes	6.2	4.34
						Husking coconuts	6.3	4.40
						Loading manure by hand	6.4	4.46
						Cutting sugar cane	6.5	4.52
						Forking	6.8	4.70
						Pedalling rickshaw without passenger	7.2	4.94
						Trimming branches of tree	7.3	5.06
						Load grain sacks— manual	7.4	5.06
						Felling tree with axe	7.5	5.12
						Hand sawing	7.5	5.12
						Pedalling rickshaw with passenger	8.5	5.72
FEMALES								
Household								
Lying	1.2	—	Walking downhill slowly	2.3	2.15	Walking with load	4.0	3.07
Sitting quietly	1.2	—	Strolling	2.4	2.22	Fetching water from well	4.1	3.13
Roasting corn	1.3	—	Squeezing coconut	2.4	2.22	Chopping wood with machete	4.3	3.25
Ironing	1.4	—	Loading earth oven	2.6	2.37	Catching crabs	4.5	3.37
Peeling sweet potatoes	1.4	—	Light cleaning	2.7	2.45	Pounding grain	4.6	3.43
Sitting sewing clothes	1.4	—	Light weeding	2.9	2.60	Walking uphill at normal pace	4.6	3.43
Podding beans	1.5	—	Sweeping house	3.0	2.67	Walking downhill fast with load	4.6	3.43
Sewing pandanus mat	1.5	—	Walking slowly	3.0	2.67	Walking uphill fast	6.0	4.27
Weaving carrying bag	1.5	—	Walking downhill at normal pace	3.0	2.67	Walking uphill with load	6.6	4.63
Preparing rope	1.5	—	Washing clothes	3.0	2.67			
Standing	1.5	—	Walking at normal pace	3.4	2.67			
Peeling taro	1.7	1.68	Sweeping yard	3.5	3.05			
Washing dishes	1.7	1.68	Moderate cleaning	3.7	3.20			
Cooking	1.8	1.70	Stirring porridge	3.7	3.20			
Squeezing coconut	1.9	1.73	Grinding grain on millstone	3.8	3.27			
Collecting leaves for flavouring	1.9	1.73	Catching fish by hand	3.9	3.35			
Breaking nuts, e.g. peanuts	1.9	1.73						
Occupational								
Spinning cotton	1.4	1.60	Machine tool industry	2.7	2.45	Sawing	4.0	3.07
Preparing tobacco	1.5	1.63	Brewery work	2.9	2.60	Binding sheaves	4.2	3.19
Picking coffee	1.5	1.63	Chemical industry	2.9	2.60	Digging holes for planting	4.3	3.25
Winnowing	1.7	1.68	Harvesting root crops	3.1	2.75	Hoeing	4.4	3.31
Office work	1.7	1.68	Furnishing industry	3.3	2.90	Digging ground	4.6	3.43
Deseeding cotton	1.8	1.70	Laundry work	3.4	2.97	Threshing	5.0	3.67
Electrical industry	2.0	1.75	Cutting fruit from trees	3.4	2.97	Cutting grass with machete	5.0	3.67
Beating cotton	2.4	1.85	Clearing ground	3.8	3.27			
			Planting root crops	3.9	3.35			

NOTE: These estimates of Energy Indices depend on an arbitrary classification of all occupations with a PAR of 1.0–2.5 being assigned to the light activity category, PAR from 2.6–3.9 to the moderate activity group and 4.0+ PAR values for specific job activities to the heavy activity group. The estimate of the length of pause varies with the three categories as listed in Table 8 of the 1985 report. In the time allocation estimates, the pauses are assumed to occupy 75 per cent, 25 per cent, and 40 per cent of the time for the light, moderate and heavy activities respectively. In this way, the Energy Indices covering the whole cost of the entire work period are derived for each work category. The PAR for the pauses is held constant and estimated at 1.54 for men, and 1.68 for women, at all three grades of work activity.

Appendix 5.1   137

1. All occupations are assigned to light, moderate, and heavy levels of activity in the following way:

    PARs 1.0–2.5 : Light ⎫
    PARs 2.6–3.9: Moderate ⎬ Categories for males and females
    PARs 4.0 +   : Heavy ⎭

2. Average estimates of PAR for pauses during specified activities are 1.54 for males and 1.68 for females. These activity ratios for pauses can be applied at all levels of activity.

3. The average length of pause time for each group was estimated as 75 per cent, 25 per cent, and 40 per cent of the time for light, moderate, and heavy activities, respectively. (Source: WHO/FAO/UNU 1985 Technical Report No. 724.)

4. Integrated energy indices are calculated for each activity by adding the specified percentage of pause time to the specified percentage of activity time in the following way:

Light Activity	75 per cent of 1.54 + 25 per cent of PAR for males
	75 per cent of 1.68 + 25 per cent of PAR for females
Moderate Activity	25 per cent of 1.54 + 75 per cent of PAR for females
	25 per cent of 1.68 + 75 per cent of PAR for females
Heavy Activity	40 per cent of 1.54 + 60 per cent of PAR for males
	40 per cent of 1.68 + 60 per cent of PAR for females

Note

ENREQ was developed before Lotus introduced Releases 3.0 and 2.2. ENREQ will not run correctly under Lotus 123 Release 3 and at the time of publication of this manual Lotus 123 Release 2.2 was not available to the authors for testing. Therefore neither version should be used for ENREQ.

Appendix 6.1 Introduction

The ENREQ (ENergy REQuirements) spreadsheet has been designed for use in calculating energy requirements of populations, based on the method proposed in the Report of the Joint FAO/WHO/UNU Expert Consultation on Energy and Protein Requirements.[1] The assumptions and principles underlying that report have been further documented in the main chapters of the present manual. It is to this manual that the ENREQ spreadsheet is intended to be a companion. Use of the ENREQ spreadsheet program is described in this User Guide.

ENREQ provides a flexible means of rapidly calculating the energy needs, according to the criteria chosen, of specific populations and population sub-groups for use both on desktop computers and portable computers in the field.

ENREQ includes recommended allowances for children, a choice of physical activity levels to be attributed to adolescents and adults, and equations for predicting basal metabolic rates (BMR) from weight. LOTUS 123 database files are provided separately for incorporation in the ENREQ spreadsheet. They contain population, weight, and height data for those FAO Member Countries that have a population greater than 300 000 (see also Appendices 1 and 2 of this manual). Provision is also made for the incorporation and storage of alternative data on population distributions, weights, heights, BMR, allowances, and physical activity levels.

Queries on the use of the ENREQ spreadsheet should be directed to the Food Policy and Nutrition Division (ESN) of FAO, Rome (telephone: 5797-6807).

Appendix 6.2 Hardware/software requirements

The ENREQ spreadsheet requires an IBM or compatible PC/XT or AT or PS/2, running under MS-DOS or PC-DOS, with at least 512 kbyte of memory.

The spreadsheet requires access to the program LOTUS 123 in order to be used. LOTUS 123 is *not* supplied with the ENREQ software. You do not need any previous experience of using LOTUS 123 in order to use ENREQ. However, if you wish to change the spreadsheet you will require greater familiarity with LOTUS 123.

The two 360 kbyte ENREQ discs contain the following files:

DISC 1		DISC 2	
	enreq.wk1		group1.pop
	program.dsk		group1.wt
	install.exe		group1.ht
			group2.pop
			group2.wt
			group2.ht
			group3.pop
			group3.wt
			group3.ht
			group4.pop
			group4.wt
			group4.ht
			fao.bmr
			fao.all
			fao.pal
			data.dsk

Step 1 The ENREQ User Guide begins at the point where LOTUS 123 has been installed in the computer. It does not attempt to duplicate information that is already contained within the LOTUS 123 manual.

In this User Guide the abbreviation ⟨CR⟩ for 'carriage return' is used. This means that the 'Return' or 'Enter' key should be pressed. The abbreviation ⟨ESC⟩ is used to mean that the ESCAPE key should be pressed.

Before using your ENREQ discs, write-protect them so that you cannot accidentally write on them. Do this by placing a sticky tab (provided along with labels in boxes of discs) over the open slot on the side of the disc. Then make back-up copies on newly formatted discs by using the MS-DOS DISKCOPY or COPY command (see your MS-DOS manual for instructions). Put the originals in a safe place and *use the copies for your work.*

Step 2 Installing ENREQ

The file 'INSTALL.EXE' on ENREQ DISC 1 should be used to install ENREQ on hard or floppy disc according to the hardware available. In order to use the install progam MS-DOS must first be loaded and drive A: selected. For floppy disc installation the INSTALL program requires two formatted 5.25″ 360 kbyte discs or one formatted 5.25″ 1.2 Mb, 3.5″ 720 kbyte or 3.5″ 1.44 Mb disc. These must be formatted using the MS-DOS FORMAT command before attempting to install ENREQ.

Installation on 360 kbyte floppy discs
Boot MS-DOS normally and place ENREQ DISC 1 in drive A: then type
    INSTALL⟨CR⟩
Select option 1 on the INSTALL menu to create the following two discs:
   PROGRAM DISC (Disc 1)
     program.dsk
     enreq.wk1
   DATA DISC (Disc 2)
     data.dsk
     group1.pop
     group1.wt
     group1.ht
     group2.pop
     group2.wt
     group2.ht
     group3.pop
     group3.wt
     group3.ht
     group4.pop
     group4.wt
     group4.ht
     fao.bmr
     fao.all
     fao.pal

Installation on 720 kbyte, 1.2 Mb and 1.44 Mb floppy discs
Boot MS-DOS normally and place ENREQ DISC 1 in drive A: or drive B: according to your disc drive configuration. Then type
    INSTALL ⟨CR⟩
Select option 2, 3 or 4 on the INSTALL menu to create an ENREQ working disc:

ENREQ DISC
　　program.dsk
　　data.dsk
　　enreq.wk1
　　group1.pop
　　group1.wt
　　group1.ht
　　group2.pop
　　group2.wt
　　group2.ht
　　group3.pop
　　group3.wt
　　group3.ht
　　group4.pop
　　group4.wt
　　group4.ht
　　fao.bmr
　　fao.all
　　fao.pal

Installation on hard disc
Boot MS-DOS normally and place ENREQ DISC 1 in drive A: then type
　　A: ⟨CR⟩
　　INSTALL ⟨CR⟩
Select option 5 on the INSTALL Menu to create the ENREQ directory on hard disc C:
　C:\ENREQ
　　program.dsk
　　data.dsk
　　enreq.wk1
　　group1.pop
　　group1.wt
　　group1.ht
　　group2.pop
　　group2.wt
　　group2.ht
　　group3.pop
　　group3.wt
　　group3.ht
　　group4.pop
　　group4.wt
　　group4.ht
　　fao.bmr
　　fao.all
　　fao.pal
If you wish to run ENREQ from a hard disc read section 3a followed by section 4. If you are running ENREQ from a floppy disc drive skip to section 3b, following it with section 4.

Step 3a　Starting up ENREQ from a hard disc
Enter LOTUS 123 then retrieve the ENREQ spreadsheet by typing
　　/FD
　　C:\ENREQ ⟨CR⟩

Then type
> */FR*

and at the prompt
> **Name of file to retrieve:**

type
> ***ENREQ ⟨CR⟩***

After approximately 30 seconds (the exact time depends on the type of computer you have) the ENREQ spreadsheet will be loaded into the computer's memory. If you are given an ERROR message flashing in the top right corner of the screen, press ⟨ESC⟩ and try again, checking that you give the correct spelling of ENREQ.

When the ENREQ title page appears on the screen and the word READY appears top right, you can activate ENREQ by pressing the ALT key, and whilst holding it down press the Z key (as instructed on your screen).

Step 3b   Starting up ENREQ from a floppy disc
Start up LOTUS 123. Place your ENREQ PROGRAM disc into your floppy disc drive. Type
> */FD*

At the prompt
> **Enter current directory: A:**

(the letter A: may be a B: on your system, depending on the disc drive from which you are running Lotus 123), type the *letter corresponding to the drive* where you have placed your ENREQ program disc (A: or B:), replacing the A: or B: offered on the screen. Then press the Return key.
Type
> */FR*

At the prompt
> **Name of File to Retrieve?**

type
> ***ENREQ ⟨CR⟩***

The ENREQ spreadsheet will take approximately one minute to be loaded into memory (the exact time depending on the type of computer). If you are given an ERROR message flashing in the top right corner of the screen, press ⟨ESC⟩ and try again, checking that you give the correct spelling. If you still get the ERROR message, check that the current directory is the one where your floppy disc has been placed.

When the ENREQ title page appears on the screen and the word READY appears top right, you can start the program. Press the ALT key, and whilst holding it down press the Z key (as instructed on your screen) in order to start the ENREQ program.

Step 4   ENREQ uses menu-driven commands which function in a similar way to LOTUS 123 itself. Therefore users familiar with LOTUS 123 should have no difficulty using these menus.

If you are new to LOTUS 123, a speadsheet is like a large ledger sheet with rows and columns, twenty rows of which can always be viewed on the screen, whereas the number of columns may differ. The fourth row of the screen downwards is dedicated to the spreadsheet whose cells are named by their position in the particular column and row in which they are situated. Columns are labelled by letters in the fourth row of the screen; rows are labelled by numbers in the left border of the screen. The position of the cursor in the spreadsheet is reported by LOTUS 123 in the top row of the screen which gives the name of the current cell. When the Main Menu is on the screen, with the word EDIT highlighted, the position of the cursor in the spreadsheet is AH1, as shown in the top row of the screen. This means that the current cell or cursor position in the spreadsheet is AH1, i.e. the intersection of column AH and row 1.

The second row of the screen is reserved for the menu. Options are listed horizontally.

The third row gives a brief explanation of the menu item currently highlighted. In order to execute an option, move the cursor with the right or left arrow key to highlight the option of choice and then press the Return key. Alternatively, to lessen the number of keystrokes required, menu options can be selected by typing the first letter of their name (you will note that within each menu, no two options begin with the same letter). Once you are familiar with the sequence of keys to be pressed, they may be pressed in rapid succession without waiting for one command to be completed before giving the next.

IN THE REST OF THIS USER GUIDE, WHEN INSTRUCTED TO CHOOSE AN OPTION FROM A MENU, EITHER HIGHLIGHT THE OPTION OF CHOICE BY MOVING TO IT WITH THE RIGHT OR LEFT ARROW AND THEN PRESS RETURN OR TYPE THE FIRST LETTER OF THE OPTION.

If you have followed the 'Getting started' section, you will now have the Main Menu on your screen which gives you the choices:

EDIT LOAD SAVE PRINT FILE GRAPH BREAK QUIT

with the word MAIN appearing in the top right-hand corner of the screen. (If you do not have this screen, hold down the ALT key and simultaneously press the Z key.)

The ENREQ program already contains the data you need in order to carry out example calculation no. 1 for which Italy has been chosen. Example calculation no. 2 uses Indian data and includes additional information on how to load other data from the databases. Italy and India have been chosen as examples because BMR equations are available for these two countries.

When in the Main Menu of ENREQ the top half of the screen is entitled 'Summary of Results' and lists all the criteria needed in order to make a calculation for energy require-ments (height is included but is only used when reference weight for height is to be calculated, since reference weights are calculated from the height data; see Command Reference sections on EDIT–RETRIEVE–WEIGHT (p. 154) and EDIT–RETRIEVE–HEIGHT (p. 155) for fuller explanations).

The lower half of the screen shows kilocalorie conversion factors. These are used to calculate the commodity equivalent in kilograms or tonnes required in order to satisfy the kilocalorie requirement. These commodity equivalents are included in the print-out of the final results.

Example calculation no. 1
STEP 1   In order to make a calculation choose the option EDIT.

The spreadsheet area of the screen will then change to show two different parts (windows) of the spreadsheet: (i) part of the Results Panel; and (ii) the first few lines of the Work Area. The Work Area is that part of the spreadsheet where the data (records) retrieved from the databases are placed so that the user can see on which data the calculations are taking place.

As you run ENREQ, information will appear in the Summary of Results part of the screen as well as in the Work Area. The results screen automatically takes data from the Work Area and makes the necessary calculations. The calculations are made so fast that the results appear in the Summary Results Area at the same time as data are retrieved from the database and placed in the Work Area.

STEP 2   The EDIT menu is now on the screen:

RETRIEVE ENTER STORE DELETE ALPHA-SORT MODE QUIT

Choose the Retrieve opion. Note that the top right corner of the screen gives you the name of the current menu.

STEP 3   The RETRIEVE menu line then appears:

POPULATION WEIGHT HEIGHT BMR ALLOWANCE LEVEL

EXTRA QUIT (This menu appears as one line on the screen)

with the word POPULATION highlighted. Information needs to be retrieved from each of the databases listed above. First retrieve population data by choosing POPULATION.

STEP 4   The menu line will now be replaced by the prompt
Name of Population:
The country name may be retrieved in one of two ways.
EITHER type it in:
Italy ⟨CR⟩
in upper or lower case or a mixture of both. Typing may be edited before pressing the

Return key. If no data are found, you made a typing mistake. Press Return and try again, checking the spelling or use the alternative method as described below.
OR type a question mark:

?⟨CR⟩

The list of countries in the population database currently in the spreadsheet will then appear on the screen. The current population database only contains Italy, which appears twice, one line for males and one for females. Move the down arrow so that the first occurrence of the word is highlighted. Press Return to retrieve both male and female population data for Italy.

Never press any key or type anything while a word appears flashing in the top right corner of the screen. This word tells you which menu is currently working and it will always appear whilst the computer is retrieving data or making calculations. You will notice various instructions flashing across the top of the screen whilst this word is showing—these are the instructions which are currently being carried out.

You should now have the name Italy and the total population figure (57 301 000) inserted in the Results area on your screen. The work area will also show the first few lines of retrieved data for males, broken down into years of age. Data have, of course, been retrieved for both sexes and all age groups, but only the first few lines of these data fit onto the screen.

STEP 5   The menu line will now be the same as in STEP 3, i.e. the RETRIEVE menu. Choose WEIGHT.

STEP 6   You will now be prompted to give information on which weight data are to be retrieved for children and adolescents

Name of weight for age curve 0–17 + (or Reference):

Retrieve actual weight data for Italy by typing in the country's name

Italy ⟨CR⟩

or by typing a **?⟨CR⟩** and using the alternative method as described for population. If you use the ? method, you will notice two entries labelled NCHS. These are explained in the Command Reference section on EDIT–RETRIEVE–WEIGHT (p. 154). You will also see an asterisk beside the country name Italy. This means that this country has its own weight and height data. See Command Reference section EDIT–RETRIEVE–WEIGHT (p. 154) for a fuller explanation.

If you make a typing mistake and you are told there is an error, press Return. You will be returned to the RETRIEVE menu. Try again.

Again, notice that the name Italy has been inserted in the Summary Results Area together with a figure for average body weight (8.3 kg). The Work Area also shows weight data for male children.

STEP 7   You now have a prompt that asks which weight data should be retrieved for adults

Name of weight for age curve 18 to > 60 (or Fogarty):

Retrieve actual weights for adults in Italy by typing

Italy ⟨CR⟩

or by using the alternative retrieval method described above by typing **?⟨CR⟩**.

If you make a typing mistake and get an error message, press Return as instructed on the screen. You will then be returned to Step 5 above, the RETRIEVE menu.

Note that the average body weight has now increased to 58.2 kg because data on adult weights have been included.

STEP 8   The option HEIGHT is included in this menu for the times when reference weights are to be looked up on the basis of actual height data (weight for height). At such times, height data must be retrieved BEFORE weight data (see Example Calculation no. 2).

In this example you do not need height data because we are calculating requirements based on weight for age (the simplest approach, described in Chapter 1.3). Either skip the HEIGHT option or choose it and at the prompt

Name of height curve (or None):

type

NONE ⟨CR⟩

STEP 9   Vital to the requirement calculation is the BMR which is derived from a set of equations based on weight. Choose BMR and you will see the prompt

Name of BMR equations:

The equations for calculating BMR based on weight, recommended in the Report of the Joint FAO/WHO/UNU Expert Consultation on Energy and Protein Requirements,[1] are known as the Schofield equations. The slope and constant values used in these equations are provided in ENREQ and are retrieved and the calculation automatically made using the previously retrieved weight data by typing

Schofield ⟨CR⟩

or by typing a *?* which will list on screen the different BMR equations provided with ENREQ. The word Schofield appears four times, twice for slope data (once for males, once for females) and twice for the constants (once for males, once for females). Any one of the four occurrences of the word may be highlighted in order to retrieve the whole set of Schofield equations. Also included in the list are the other BMR equations provided with ENREQ. These are explained in Command Reference section on BMR (p. 156).

STEP 10   The next option from the menu to be chosen is ALLOWANCE. The prompt will appear

Name of allowance profile:

Here the decision must be made as to what kilocalorie intake per kilogram of body weight to use for children aged 0–9+. The choices may or may not include infection allowances and/or catch-up growth. The choices are summed up with the following names where DC and LDC stand for developed and less developed countries respectively:

Requirement
Desirable
Infection
DC
LDC
Urban DC
Urban LDC
Rural DC
Rural LDC

For this example choose Desirable EITHER by typing in the word

Desirable ⟨CR⟩

OR by typing

? ⟨CR⟩

and highlighting one of the two occurrences on the screen of the chosen allowance. This will retrieve the allowances for both males and females.

Note that the Work Area and Summary Results are gradually filling up with more data. (See Command Reference Section on EDIT–RETRIEVE–ALLOWANCE (p. 156) for an explanation of choices available.)

STEP 11   Physical activity levels (PALs) for adolescents and adults may be chosen next. Choose the option LEVEL from the Retrieve Menu (the word LEVEL is used to indicate physical activity LEVEL). The prompt will appear

Name of Physical Activity Level profile:

One of the following must be chosen:

LDC

Urban LDC
Rural LDC
DC
Urban DC
Rural DC
Desirable/L
Desirable/M
Desirable/H
Requirement/L
Requirement/M
Requirement/H

where L, M, and H stand for low, moderate and high respectively.

For this example, choose **Desirable/M**, either by typing the word or by typing a *?* and choosing from the screen list. (Each choice is listed twice, once for males and once for females. Highlighting one of the two will retrieve both.) (See Command Reference Section on EDIT–RETRIEVE–LEVEL (p. 157) for an explanation of the available choices.)

STEP 12   Choose the last component of the calculation, EXTRA, which is the extra allowance for pregnant women. Three choices are listed in a sub-menu as follows:

ACTIVE   SEDENTARY   REQUIREMENT

These are discussed in section 5.10 of the present manual. For our example, choose **REQUIREMENT**.

The results area of the screen should now be complete, showing you all the choices you made and giving you the results of the calculation. The last result listed, the PER CAPUT ALLOWANCE, will give you the per caput kilocalorie average requirement for the population chosen (including children of course) according to the criteria selected and should read 2381.

Total population and pregnancy requirements are given in thousands (ten to the power of 3) of calories, as there is not enough room to give the entire figure.

Although it cannot be viewed from this screen, the work area is also complete.

STEP 13   To print the results, exit the Retrieve Menu by choosing QUIT. The previous menu, the Edit Menu is then shown. Exit this menu also by choosing QUIT.

You will then find yourself in the Main Menu where one of the options is PRINT. Make sure your printer is switched on and on-line then choose PRINT. The print-out lists the data used in the calculations, the summary results and the commodity equivalents. If you do not have a printer attached to your computer, giving the print command could result in your computer refusing to respond to any command at all. Should this happen, reboot (i.e. start again) your computer by pressing the CTRL–ALT–DEL keys simultaneously or by switching the computer off and on again after a few seconds. Reload ENREQ and start again.

STEP 14   Try using different allowances and physical activity levels to see how these affect your results. Try as many combinations as you wish. You may also wish to use Italian BMR equations by typing

Italy ⟨CR⟩

at the BMR prompt.

Consult Chapter 5 of the manual and the Command Reference section of this User's Guide for a complete discussion of the options and their effects.

STEP 15   Continue to example calculation no. 2 or quit ENREQ and LOTUS 123 by choosing QUIT in the Main Menu. When asked

Save current worksheet (y/n)?

Type
N⟨CR⟩
This will NOT save your calculation. See Command Reference sections on SAVE (p. 163) and on FILE (p. 164) for a discussion of when you should save your spreadsheet or file your calculations to a print file.

Example calculation no. 2
If you have followed example calculation no. 1 or the Getting started section of this User Guide, your screen should now show the Main Menu. If this is not the case, load ENREQ and then press the ALT and Z keys simultaneously to activate the program.

In this second example calculation, we will retrieve population, weight, and height data for a country which is not in the current databases stored in the ENREQ spreadsheet (which at the moment only includes population, weight, and height databases for the example calculation no. 1, i.e. Italy).

Data will be loaded into the spreadsheet from the disc if you are working from a floppy disc or from the hard disc if you have installed both program and data discs on your hard disc.

STEP 1a For floppy disc users
Choose the LOAD option and at prompt
Insert DATA disc then press Return
place your copy of ENREQ disc no. 2 (which is now labelled DATA DISK) in your floppy disc drive, replacing the ENREQ PROGRAM DISK. Close the drive door and press Return.

STEP 1b For hard disc users
Choose the LOAD option.

STEP 2 The LOAD menu now on screen gives the name of the various databases which can be loaded into memory:
**POPULATION WEIGHT HEIGHT BMR ALLOWANCE LEVEL
EXTRA QUIT** (This menu appears as one line on the screen)

STEP 3 The only databases which need to be loaded at this stage are POPULATION, WEIGHT, and HEIGHT. (The BMR, ALLOWANCE, and LEVEL databases provided with ENREQ are already resident in the ENREQ spreadsheet and the options are included in this menu only because the user is allowed to create additional BMR, ALLOWANCE, and LEVEL databases using other options of ENREQ and these may be LOADed from this menu. For a full explanation of this command, see the Command Reference section on LOAD, p. 161.)

In our example we will load population, weight, and height data for India. Choose POPULATION first. You will then be asked
Name of population database to load:
For reasons of memory space limits, the databases on countries' populations, weights and heights have been divided into regional groups. India is in the Group2 database (regional groupings are listed in Command Reference section on LOAD, p. 162, and the countries in each group are listed on p. 166). Therefore in answer to the prompt, type
Group2 ⟨CR⟩

STEP 4 When the word LOAD stops flashing in the top right corner of your screen, you are returned to the same menu as in STEP 2. Repeat STEP 3 for WEIGHT and HEIGHT, LOADing Group2 weight and height data.

STEP 5 Quit the LOAD menu by choosing QUIT.

STEP 6 Population, weight, and height databases for Group2 have now been loaded from disc and stored in the ENREQ spreadsheet. The data (records) for India must be retrieved from the databases and placed into the Work Area. The RETRIEVE command is a sub-command of the EDIT menu. Therefore choose EDIT from the Main Menu.

STEP 7   Choose RETRIEVE from the EDIT menu.

STEP 8   Retrieve POPULATION data first. Type
   India ⟨*CR*⟩
or *?* ⟨*CR*⟩. The question mark will list the GROUP2 population database on the screen. Choose India by moving the down arrow to either of the 2 occurrences of the word and press Return. This will retrieve both male and female population data for India.

   If you did example calculation no. 1 in this session at your computer, the name Italy in the Results Area will now be replaced by India against the word POPULATION.

STEP 9   In this example we demonstrate the use of reference weights for height using Indian HEIGHT data rather than using Indian actual weight data. To do this, the HEIGHT data must be retrieved first so that ENREQ can use them to retrieve the reference weight for height from look-up tables. For this reason the HEIGHT data must be retrieved before the WEIGHT data. Retrieve HEIGHT data for India in the same way as you retrieved Population data.

STEP 10   Next retrieve WEIGHT data by choosing WEIGHT on the menu. To retrieve reference weight for height data for children and adolescents, at the prompt type
   Reference ⟨*CR*⟩
You may not type in a *?* but only the word Reference.

STEP 11   At the prompt for adult weight data, retrieve reference weight for height data for adults by typing
   Fogarty ⟨*CR*⟩
Make sure you get the spelling correct as you cannot retrieve this word from the screen by typing a *?*. At this stage the interim results in the Results Panel give an average body weight of 41.4 kg.

STEP 12   If you previously did example calculation no. 1 *without having retrieved data other than those in the tutorial exercise,* your total per caput requirement figure will now read 2046 kilocalories. Otherwise retrieve BMR, ALLOWANCE, LEVEL, EXTRA data as in example calculation no. 1.

STEP 13   Print as per instructions in example calculation no. 1.

STEP 14   Try various combinations of allowances, physical activity levels, and weight to see how these affect the results. An Indian set of equations for calculating BMR is available by typing *India* ⟨*CR*⟩ at the BMR prompt. Again consult Chapter 5 of the manual and Command Reference sections on EDIT–RETRIEVE–ALLOWANCE (p. 156) and EDIT–RETRIEVE–LEVEL (p. 157) for a full discussion of PALs and allowances.

ENREQ's menu structures are shown in two columns in the following tree diagram. Each command of each menu is explained thereafter in the order of the diagram.

Edit
    Retrieve
        Population
        Weight
        Height
        BMR
        Allowance
        Level
        Extra
            Active
            Sedentary
            Requirement
        Quit
    Enter
        Titles
        Data
        Conversion
        BMR
        Quit
    Store
        Population
        Weight
        Height
        BMR
        Allowance
        Level
        Quit
    Delete
        Population
        Weight
        Height
        BMR
        Allowance
        Level
        Quit
    Alpha-sort
        Population
        Weight
        Height
        BMR
        Allowance
        Level
        Quit

(*cont.*)
    Mode
        Normal
        Split
        Reset
        Clear
        Erase
            Population
            Weight
            Height
            BMR
            Allowance
            Level
            Quit
        Quit
    Load
        Population
        Weight
        Height
        BMR
        Allowance
        Level
        Quit
    Save
        Entire
        Database
            Population
            Weight
            Height
            BMR
            Allowance
            Level
            Quit
    Print
    File
    Graph
        Population
        Weight
        Height
        Individual
        Total
        Save
        Quit
    Break
    Quit

INTRODUCTION

Before explaining each ENREQ command one by one, a general description is necessary to better understand the different levels of data usage and storage in ENREQ.

LOTUS 123 spreadsheets are loaded into the PC's memory for execution. This means that any calculations or additions made to the spreadsheet are only stored in the spreadsheet in memory, not on disc. They will be retained only as long as you remain in ENREQ. To have the latest version of the spreadsheet, with additions and calculations, written to disc, the spreadsheet must be saved to disc with the SAVE command. If the SAVE command is not used, the disc version of the spreadsheet will remain as it was before you worked on it.

Because of memory limits, not all the databases provided together with ENREQ can be stored in a LOTUS 123 spreadsheet, and consequently ENREQ, at the same time. It is for this reason that the databases for population, weight, and height have been split up into four different country groupings which, loaded one at a time, do not exceed the memory limits (see also section on LOAD, p. 161).

When using ENREQ with data from the databases provided on the ENREQ discs (or already installed on the hard disc), the databases for population, weight, and height for the country group to which the country in question belongs must be first LOADed into the ENREQ spreadsheet. Once LOADed, they can be viewed and data (records) retrieved (using EDIT–RETRIEVE) from one database at a time. Retrieved data are inserted into the appropriate part of the Work Area and at the same time the necessary calculations are made by the program and results shown in the Results Panel.

When a user wishes to use his/her *own* data, these may be ENTERed into ENREQ manually. The ENTER command will let you enter data into the Work Area of the spreadsheet. Since data in the Results Panel are automatically updated when data are introduced into the Work Area, the Results Panel is immediately affected when data are ENTERed by the user. However, ENTERing data into the Work Area does NOT store them into the database, nor write them to disc.

If ENTERed data are to be used more than once in one computer session, they should be STOREd to the appropriate database in ENREQ using the STORE command. Once new data are STOREd in a database in ENREQ they may be RETRIEVEd later in the same computer session.

Since ENREQ is a LOTUS 123 spreadsheet which is loaded into and then accessed by the computer's memory, the newly ENTERed and STOREd data must also be SAVEd to disc if needed in a future computer session, using the SAVE command (see section on SAVE, p. 163).

In order to avoid LOADing databases into ENREQ at each computer session, a short-cut may be taken by users who *regularly* use the same data sets, e.g. one or more countries from the same regional grouping or user-created data. First LOAD into ENREQ the country group databases for population, weight, and height of choice and then SAVE them with the ENREQ program itself (see section on SAVE). Subsequently, each time ENREQ is activated, the data sets of choice will be loaded into memory at the same time as the program.

Since the ENTER and STORE commands do not write to disc, the only time you need to use a disc (in a floppy disc system) is for the SAVE command.

Data (records) that have been STOREd to databases may be DELETEd from the database currently in ENREQ using the DELETE command. This will erase the entry from the database in ENREQ but not from the disc (if it had been saved to disc in the first place!). To erase that entry also from the disc version of the database, the database should be reSAVEd once the DELETion of the item has taken place, thus resulting in the former database being written over by the newer one. Note that the entry is *not* deleted from the Work Area. (See section on EDIT–DELETE, p. 160).

Figure 1 summarizes the previous discussion.

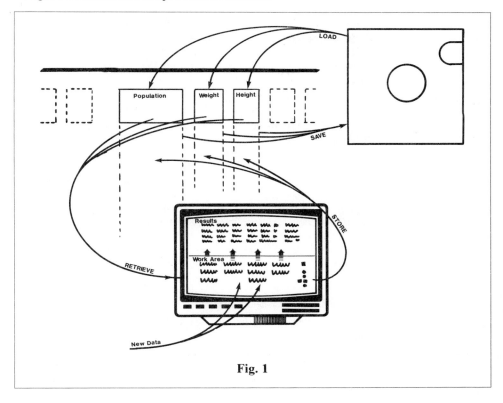

Fig. 1

COMMANDS

The Main Menu is reached by pressing the ALT and Z keys simultaneously and is
    EDIT LOAD SAVE PRINT FILE GRAPH BREAK QUIT

EDIT

The EDIT Menu is as follows:
    RETRIEVE ENTER STORE DELETE ALPHA-SORT MODE QUIT

EDIT-RETRIEVE

Choosing RETRIEVE gives the following sub-menu:
    POPULATION WEIGHT HEIGHT BMR ALLOWANCE LEVEL
    EXTRA QUIT (This menu appears as one line on the screen)

EDIT–RETRIEVE–POPULATION

United Nations' 1985 population estimates for FAO Member Countries with populations greater than 300 000 may be retrieved using this option. Population numbers are given by age and sex groups relevant for calculation of population energy requirements. Total numbers of pregnant women (calculated as 75 per cent of total annual births) are also provided. See Appendices 1.1 and 1.2 for population numbers used in ENREQ. **(N.B. Before data can be RETRIEVEd, they must be present in the databases in ENREQ. If this is not the case, the appropriate database must be LOADed into ENREQ from disc—see command LOAD.)**

At the prompt

    Name of population:

EITHER type in the ***name of the selected country*** $\langle$***CR***$\rangle$. Check the spelling/abbreviation of the country name against the list provided on p. 166. Typing may be edited before pressing Return.

OR type a question mark **?** $\langle$***CR***$\rangle$. This will cause the names of countries in the populuation database currently in ENREQ to be listed on the screen, with two lines of data for each country (one for males and one for females). Make your choice by moving the down arrow until it highlights one of the two occurrences of the name of the country of choice and then press Return. You may exit from this screen without making a choice by moving the cursor to highlight the dotted line before the first name in the list and pressing Return.

A previously retrieved country population may be retained by pressing Return at the prompt.

EDIT–RETRIEVE–WEIGHT

Weight data, broken down by age and sex categories, have been collated by FAO for a number of countries but not for all of those included in ENREQ's population database. When complete data do not exist, data for a country with a similar weight structure have been substituted. To see which countries have complete data and for which countries these data are used as substitute data, refer to Appendices 2.1, 2.2, and 2.3. A single available adult value has been used for all adults except in those few cases where complete data are available for the different adult age groups.

In addition to databases containing *actual* weights by country, ENREQ contains a series of lookup tables which provide reference weight values according to sex, age group, and height. These are commonly accepted as references for international comparisons. (*These reference weights can only be accessed once height data have been retrieved.*) The use of these reference values allows a normative approach to calculating energy requirements (see sections 5.2, 5.5–5.9 and Table 5.13 of manual). The reference weights are:

NCHS   U.S. National Center for Health Statistics (NCHS) *weight for age* reference values for ages 0–17+ (see Appendix 3.5)

Reference   U.S. National Center for Health Statistics (NCHS) *weight for height* reference values for ages 0–9+ (see Appendix 3.1)

           Baldwin desirable *weight for height* reference values for ages 10–17+ (see Appendices 3.2 and 3.3)

Fogarty   Acceptable *weight for height* values for adults (18–>60 years) of a North American population based on 1960 Metropolitan Life Insurance Tables (see section 5.5 and Appendix 3.4) and adopted at two conferences on obesity during the 1970s held at the John E. Fogarty International Centre Conference on Obesity.

You will first be prompted to retrieve weight data for children and adolescents aged 0–17+ years

    Name of weight curve for age 0–17+ (or Reference):

You may retrieve one of three different types of children's weight data:

  (i) To retrieve children's *actual weights*:

EITHER type in the **name of the country** ⟨*CR*⟩ as listed on p. 166 (if the selected country does not have its own data, the substitute data will be automatically retrieved)

OR type a *?* ⟨*CR*⟩ for the screen to list the current children's weight database. If the current weight database is empty, the database of choice must be LOADed from disc into the ENREQ spreadsheet using the LOAD command.

Using the latter method, use the down arrow to move to the name of the country whose data are to be retrieved and press Return. Each country is listed twice, once for males and once for females. Highlighting either occurrence of the country name of choice will retrieve both male and female children's weights. You may not use ⟨ESC⟩ to exit from this screen. The only way to exit without choosing a country is to move to the dotted line above the first name in the list and press Return. You will notice that some countries have an asterisk at the left of their name. These are the countries for which FAO has collated complete data sets (see above). A '+' sign beside a country name indicates that weighted averages have been calculated using data from more than one study.

(ii) To retrieve U.S. National Center for Health Statistics (NCHS) *reference weight for age* data, type

    NCHS ⟨*CR*⟩

or type a *?* ⟨*CR*⟩ and choose NCHS from the screen list.

(iii) To retrieve *reference weight for height* data which includes the NCHS reference weight for height data for children aged 0 to 9+ years and the Baldwin reference weight for height data for adolescents aged 10 to 17+ years, type

    REFERENCE ⟨*CR*⟩

Please note that Reference weight for height data may not be chosen by typing a ?. Pressing Return at the prompt will retain the previously used choice.

You will then be prompted to retrieve adult weight data

    Name of weight for age curve 18–>60 (or Fogarty):

You may retrieve one of two different types of adult weight data:

(i) To retrieve *actual adult weights*, type the **name of the country** ⟨*CR*⟩ as given on p. 166 or type a *?* ⟨*CR*⟩ and select it from the screen which will list the current adolescent and adult weight data-base.

(ii) To retrieve *reference weight for height* data for adults type the word

    FOGARTY ⟨*CR*⟩

Note that Fogarty reference weight for height data may not be chosen by typing a ?. Previously retrieved adult weight data may be retained by simply pressing Return at the prompt.

N.B. REFERENCE and FOGARTY weight for height data use actual HEIGHT data from the country in question. Therefore REFERENCE weight for height data for children and FOGARTY weight for height for adults may be retrieved only after *height data have been retrieved. The reference weights are looked up in their respective tables and although not inserted into the Work Area, are printed in the Results Page.*

EDIT–RETRIEVE–HEIGHT

The height data must be retrieved before weight data if reference weight for height weights are to be used. As already stated above for weight data, the height database does not contain original data for every country and therefore substitute data are sometimes automatically retrieved—refer to Appendices 2.1, 2.2 and 2.3 for explanations.

At the prompt

    Name of height curve (or None):

type the **name of the country** ⟨*CR*⟩ as given on p. 166 or a *?* ⟨*CR*⟩ in order to choose from the database listed on screen. If the height database is empty or does not contain the database you require, LOAD the appropriate database from disc using the LOAD command.

If height data are not required because actual weights or children's weight for age data are used, either skip this option or type

NONE ⟨CR⟩

EDIT–RETRIEVE–BMR

Normally the BMR quadratic equations used to calculate energy requirements are those contained in the FAO/WHO/UNU Technical Report on Energy and Protein Requirements[1] which are known as the SCHOFIELD[2] equations. However, local quadratic equations may also be used. Sample equations are included in ENREQ for INDIA and ITALY as well as a set of equations called CONSTANT. The set of equations called CONSTANT are really an abbreviated version of the Schofield equations. The CONSTANT equations assign the Schofield equations for the age group 18–29+ to all adult age groups, thus giving a constant BMR throughout adulthood which could prove useful if constant weights for all adult age groups are being used. Other BMR quadratic equations may be added—see sections on EDIT–ENTER–BMR (p. 160) and EDIT–STORE (p. 160).

At the prompt

Name of BMR equations:

you will usually type *SCHOFIELD* ⟨CR⟩ to retrieve the Schofield equations. The only other equations supplied with ENREQ may be retrieved by typing *INDIA* ⟨CR⟩ or *ITALY* ⟨CR⟩ or *CONSTANT* ⟨CR⟩. The Indian and Italian equations are derived from actual weights and measured BMR data from relatively large sample populations of these two countries.

To view the various sets of equations, type a *?* ⟨CR⟩ at the above mentioned prompt. When viewing the equations without wishing to retrieve a set, exit the list by moving the cursor to highlight the dotted line above the first name in the list. To retrieve a set of equations, move to the name required. The name is repeated four times. Any one of the four occurrences may be highlighted in order to retrieve the whole set of equations.

EDIT–RETRIEVE–ALLOWANCE

The allowance option assigns a per kg body weight energy value in order to calculate the requirement for children age 0–9+ years.

The choices are discussed thoroughly in sections 5.1 to 5.4 of the manual. For children aged 0–9+ years, the FAO/WHO/UNU Report[1] begins with a requirement level based on the energy intake level of apparently healthy but possibly underweight children, thus maintaining the status quo. This REQUIREMENT level does not allow for catch-up growth or provide any extra allowance for the weight loss resulting from repeated infections which often affect children in less developed countries. To this REQUIREMENT value one can add an additional 5 per cent allowance for assuring desirable physical activity. This has been done in ENREQ and the resulting allowance is labelled DESIRABLE. A third level, labelled INFECTION, uses the DESIRABLE allowance with an additional energy allowance for 0–2+ year olds to allow for catchup growth of weight lost during intermittent infectious illnesses.

Six other allowance options are offered but all are based on choices discussed above. The complete set is as follows:

Allowance (kcal per kg, 0–9+ years)	Explanation	Comment/clarification
REQUIREMENT	For maintenance of status quo	Energy need based on existing weight

Allowance (kcal per kg, 0–9+ years)	Explanation	Comment/clarification
DESIRABLE	Energy REQUIREMENT with an additional allowance of 5%	Allows for desirable physical activity
INFECTION	DESIRABLE allowance with an additional extra allowance for 0–2+ years	Allows for desirable activity and recovery from infection in infants
LDC URBAN LDC RURAL LDC	as INFECTION above	
DC URBAN DC RURAL DC	as DESIRABLE above	

The values of the options described above are explained in Tables 5.1, 5.7, and 5.8 and can be viewed and retrieved by using EDIT–RETRIEVE–ALLOWANCE and typing *?* ⟨*CR*⟩ at the prompt

Name of allowance profile:

Move the cursor to the option of choice and press Return. If you are only viewing and do not wish to retrieve, move the cursor to the dotted line above the first name in the list and press Return to exit.

If you do not need to view the choices, you may type in the ***name of the allowance*** ⟨*CR*⟩ of your choice at the prompt. If you have created your own allowances, these may be retrieved at this stage using the name you created, provided the correct database is currently stored in ENREQ.

EDIT–RETRIEVE–LEVEL

In order to calculate requirements, the actual or desirable *Physical Activity* LEVEL (PAL) for the adolescent ages from 10–17+ and adult age groups 18–59+ and >60 should be known or estimated.

To view and/or retrieve a PAL type *?* ⟨*CR*⟩ at the prompt

Name of Physical Activity Level profile:

If you do not need to view the PAL list, type the ***name of the PAL*** ⟨*CR*⟩ of choice at the prompt. The choices are briefly described below:

PAL	Description/explanation	Comments	Reference
LDC	Less developed country comprises 80% rural and 20% urban populations	Rural assumes 75% moderate + 25% heavy activity levels. Urban assumes 50% light + 50% moderate activity levels	Table 1.8, p. 26
DC	Developed country comprising 70% urban and 30% rural population	Urban assumes 75% light activity + 25% moderate activity levels. Rural assumes 100% moderate activity	Table 1.8, p. 26

PAL	Description/explanation	Comments	Reference
URBAN DC	Urban developed country	Assumes 75% light + 25% moderate activity levels	
RURAL DC	Rural developed country	Assumes 100% moderate activity	
URBAN LDC	Urban less developed country	Assumes 50% light + 50% moderate activity levels	Table 4.1, p. 59
RURAL LDC	Rural less developed country	Assumes 75% moderate + 25% heavy activity levels	
DESIRABLE/L	Desirable light activity level	Light requirement with desirable activity allowance	
DESIRABLE/M	Desirable moderate activity level	Moderate requirement with desirable activity allowance	
DESIRABLE/H	Desirable heavy activity level	Heavy requirement with desirable activity allowance. Note: this is only likely to be applicable to a few population subsets who engage in high activity levels during work or exercise periods	Table 3.5, p. 54 Table 5.8, p. 74
REQUIRE- MENT/L	Light activity at requirement level		
REQUIRE- MENT/M	Moderate activity at requirement level		
REQUIRE- MENT/H	High activity at requirement level	Note: this is only likely to be applicable to a few population subsets who engage in high activity levels during work or exercise periods	

EDIT–RETRIEVE–EXTRA

As extra allowances are needed for pregnant women, these must be taken into account in the calculation. ENREQ gives three choices of pregnancy allowances, with the following names:

ACTIVE   SEDENTARY   REQUIREMENT

These are discussed at the end of Chapter 5 in section 5.10 of the present manual. Choosing ACTIVE will give an extra allowance of 285 kcal per caput per day to the estimated number of pregnant women; SEDENTARY will give 200 kcal per caput and REQUIREMENT 100 kcal per caput per day.

The three choices are presented in a sub-menu to facilitate their retrieval. The number of pregnant women in each country is supplied by ENREQ and is automatically retrieved when population data are retrieved. The number has been calculated as 75 per cent of the annual number of births, as given in Appendix 1.2.

EDIT–ENTER

The ENTER commands provide a means of entering new data directly into the work area of the spreadsheet. The numbers in the Results Panel are automatically updated as new data are entered. Note, however, that newly entered *height* values are only used in the calculations if REFERENCE or FOGARTY weights are selected on the results panel.

If you have not already done so, it is advisable to read the introductory section of the Command References before using the ENTER command.

Data can be entered in four separate areas as shown in the sub-menu:

TITLES DATA CONVERSION BMR QUIT

The four different ENTER options are described below.

EDIT–ENTER–TITLES

This command allows you to alter the titles used to label data when printing out results. It also lets you give labels to new values which you are going to ENTER into the Work Area. This option is useful when creating new data entries but it is important to remember that the titles simply label the current data in the work area which may be the values previously retrieved from the database for another calculation. Type in the new title/label(s). Return to the EDIT–ENTER menu by pressing Return. Changing the label does not have any effect on the data. In order to enter new data use the EDIT–ENTER–DATA option.

EDIT–ENTER–DATA

This is the command which enables you to enter your own data. It is highly recommended that you use your own population, weight, height, and PAL data where possible. You may also wish to select only certain age/sex groups of a country's population (see next paragraph), or you may work on population sub-groups e.g. a certain region, institution or age group for which you have your own data.

Choosing the EDIT–ENTER–DATA option takes the cursor to the bottom half of the screen, the Work Area, into the column entitled Selector. The default value in this column is 1, which means that the corresponding age/sex group is *included* in the calculation. The only other valid value allowed in this column is zero, which will *exclude* the line from the calculation. This option is useful in defining sub-groups of populations where only certain age/sex groups are to be included.

The columns where your own data may be added are POPULATION, WEIGHT, HEIGHT, ALLOWANCE (for 0–9+ age group) and PAL (for adolescents and adults). When typing in numerical data, do not type in the commas as they are provided automatically by ENREQ. Do not type leading blanks (by pressing the space bar) as these will be interpreted as letters rather than numbers and will give an error in the calculations. Return to the EDIT–ENTER menu by pressing Return.

Experienced LOTUS 123 users may wish to enter data into a LOTUS 123 spreadsheet outside ENREQ and then LOAD them as a database. Use the same format as the provided databases.

EDIT–ENTER–CONVERSION

ENREQ offers the option of expressing the per caput allowance in terms of the quantities of five different commodities or grains. Appropriate information has been supplied in the spreadsheet for expressing the per caput allowance as whole grain wheat, maize and millet, and milled rice. These grains have been selected because they are commonly used to express national food needs as grain equivalents. You should be aware that if you convert the commodity from whole grain to flour or some other processed product or vice versa, you must allow for the milling or extraction rate. Thus, for example, 0.55 kg of milled rice originates from 0.83 kg of paddy rice that has been milled to 67 per cent of its original weight.

The commodities and their conversion factors may be edited or replaced using this command. Return to the EDIT–ENTER menu by pressing Return.

EDIT–ENTER–BMR

The regression equations used to predict basal metabolic rate from body weight require slope and constant data for the age–sex groupings 10–17+, 18–29+, 30–59+, and > 60. If you have access to calorimetric measurements for the population you are dealing with, and you have calculated your own quadratic regression equations, these may be entered here. Return to the EDIT–ENTER menu by pressing Return.

EDIT–STORE

New data which has been ENTERed into the Work Area may be STOREd into the data-bases currently in ENREQ with this command. Data must be STOREd in one database at a time.

Choose to which database you wish to STORE data (a record) from those listed in the menu:

POPULATION WEIGHT HEIGHT BMR ALLOWANCE LEVEL QUIT
Unless a title/label has already been ENTERed into the summary results panel using EDIT–ENTER–TITLE, you will now be prompted for one. If the label you use is already listed in the database, you will be told that it already exists and asked if you want to replace it:

. . . data already exists: replace (y/n)?
A negative answer will bring you back to the EDIT menu. STOREd data (records) are added to the end of the database of choice currently in ENREQ. If new STOREd data are to be used in a future computer session, the database in which they have been STOREd should also be SAVEd to disc (see section on SAVE, p. 163). STOREd data may be viewed and RETRIEVEd (EDIT–RETRIEVE), DELETEd (EDIT–DELETE) and SORTed (EDIT–ALPHA–SORT).

EDIT–DELETE

Database entries (records) can be DELETEd from databases currently in ENREQ with this command. The following menu is given

POPULATION WEIGHT HEIGHT BMR ALLOWANCE LEVEL QUIT
First tell ENREQ from which database you wish the entry to be DELETEd. Then give the title/label of the entry which is to be deleted (or type *?* ⟨*CR*⟩ to see the database listed on screen and choose by moving down the arrow to highlight the entry to be deleted). If you change your mind about wanting to DELETE an entry, instead of typing the label or typing a ?, press the Escape key ⟨*ESC*⟩ or simply press Return to abandon this command.

You will notice that the Work Area is not affected when an entry is DELETEd. This is because the entry is DELETEd from the database not from the Work Area. If after a DELETion, the database or the entire spreadsheet is saved to disc using the SAVE command, the database will not contain the deleted entry the next time it is loaded.

EDIT–ALPHA–SORT

This command sorts the chosen database alphabetically. Since new entries are added to the end of the appropriate database, records are no longer in alphabetical order. This command will sort the database and place the entries in their correct alphabetical position, for easier reference.

EDIT–MODE

This allows you to change the screen mode as well as reset and clear parts of the Work Area and databases. The sub-menu is:

FULL   SPLIT   RESET   CLEAR   ERASE   QUIT

EDIT–MODE–FULL

ENREQ usually splits the screen into two separate windows in order to show results as well as the top part of the Work Area. If you prefer to have the whole screen dedicated to the Work Area, choose this option.

EDIT–MODE–SPLIT

This is the default, a split screen mode where two windows are displayed simultaneously. Use this option to return to the default if you have previously used the full screen mode.

EDIT–MODE–RESET

Resets any sub-group selectors in the work area back to 1 so that all age groups are included in calculations.

EDIT–MODE–CLEAR

Clears the current data in the ENREQ Work Area except for the sub-group selectors and commodity conversion values.

EDIT–MODE–ERASE

This command erases the data from databases in memory, one at a time. You will be prompted to say which set of data is to be erased. Use this command to empty the databases if you wish to STORE newly entered data to a new empty database.

LOAD

If you have not already done so, it is recommended that you read the introductory section of Command References before using LOAD.

The original ENREQ spreadsheet contains population, weight, and height databases for Italy, the country used in the first tutorial example, as well as FAO databases for BMR, ALLOWANCEs and PALs. The databases for all other countries are provided on the second ENREQ disc and can be installed on your hard disc (see Getting Started). Since LOTUS 123 spreadsheets run in memory, all of the spreadsheet information must fit into memory. For reasons of limits in memory space, not all data for all countries can be stored in ENREQ at once. Therefore, once the ENREQ program has been activated, the required country group databases will have to be LOADed into ENREQ from disc (unless the user has previously SAVEd ENREQ with the required databases already LOADed).

Having chosen the LOAD option, users working from a floppy disc will be prompted to insert the disc containing the desired data and to press Return. This message will not appear for users who have installed ENREQ for a hard disc system (see Getting Started).

For users familiar with LOTUS 123, the LOAD command combines an external database from the disc with the ENREQ spreadsheet in memory. The external database files have exactly the same arrangement of data as those used in the ENREQ spreadsheet. In fact, a quick way to enter your own data into databases is to do it directly in a LOTUS 123 file, using the ENREQ databases as models.

Databases which can be loaded into memory are listed in the LOAD menu as follows:

POPULATION WEIGHT HEIGHT BMR ALLOWANCE LEVEL QUIT

The databases for population, weight, and height have been divided up into four regional groups as they will not all fit into memory at the same time. See Appendix 2.2 or p. 166 for a list of countries in each FAO regional group. The groupings of regions for ENREQ are:

Africa and Near East	GROUP1
Asia and Southwest Pacific	GROUP2
Europe and N. America	GROUP3
Latin America	GROUP4

When loading POPULATION, WEIGHT, and HEIGHT databases, check to which group the country to be retrieved belongs and type the name of that group at the prompt.

Name of .... database to load:

(The dots represent the word on screen which tells you which type of database you are loading.)

If you make a typing mistake, ENREQ will inform you of an error and tell you to start again by pressing ALT and Z simultaneously. If you inadvertently erase the FAO BMR, Allowance or PAL databases, these can be reLOADed from your hard disc or the appropriate data disc. Choose the appropriate database from the LOAD menu and then type

FAO ⟨CR⟩

If you wish to run the tutorial example no. 1 after having LOADed other databases, or after having SAVEd the whole ENREQ spreadsheet with other databases LOADed, you must go back to your original ENREQ program or its copy.

Should it be necessary to perform repeated calculations for the populations from a set of countries from different FAO regions a 'working' database can be constructed as follows. Example: To construct a working database containing countries X, Y, and Z from GROUP1, GROUP2, and GROUP3 databases respectively.

STEP 1   LOAD GROUP1 population, weight, and height databases.

STEP 2   RETRIEVE country X population, weight, and height records to the Work Area.

STEP 3   Use the MODE–ERASE command to empty the GROUP1 population, weight, and height databases.

STEP 4   STORE country X records to the new empty population, weight, and height databases.

STEP 5   SAVE newly created databases to disc with a new name, e.g. WORKING.

STEP 6   Repeat STEPs 1 to 3 for GROUP2 and country Y.

STEP 7   LOAD WORKING population, weight, and height databases.

STEP 8   STORE country Y population, weight, and height records to their respective databases (which will add country Y to the WORKING databases).

STEP 9   SAVE WORKING databases to disc.

STEP 10   Repeat STEPs 6 to 9 for GROUP3 and country Z.

This is a very laborious method which is suitable when only very few countries have to be put into a new database. However, a more efficient way to create a new database is to use LOTUS 123 commands. Exit ENREQ and in another worksheet use /FILE

COMBINE to put all the group databases of one type (e.g. weight) together. Use /FILE EXTRACT to reduce the size of the database to those countries with which you want to work. Save your new database as a worksheet and LOAD it into ENREQ when needed. Although this way is more efficient, it does require a knowledge of LOTUS 123.

SAVE

Alterations to the work area of the spreadsheet and/or additions STOREd in the databases will be lost at the end of your computer session unless the spreadsheet or database is saved to disc.

The choice of saving the entire spreadsheet or individual databases is given with the sub-menu:

ENTIRE   DATABASE

SAVE–ENTIRE

The entire spreadsheet may be saved to disc, replacing the version of ENREQ previously on disc. *Only* use this option if you want to save ENREQ with the current databases loaded and with the changes and/or additions you have made to it during your computer session. If you have made additions and/or deletions to certain databases, or created new databases, use the SAVE–DATABASE option to save these databases to disc.

SAVE–DATABASE

This command is useful if you have made additions or deletions to a database. When SAVing the new version of the database to disc, you will be prompted for the type of the database you want to save from the following:

POPULATION WEIGHT HEIGHT BMR ALLOWANCE LEVEL QUIT
(If you are using a floppy disc system, you will be prompted to insert a disc before proceeding to save the database.) You will next be prompted for the name of the database to be saved. If a database file with the same name already exists on the disc it will be overwritten. This implies that if you have added an entry to a certain database in memory by STORing it, that you should use that same database's name when saving it. But should you wish to save this database with a new name, give the new name here.

If the disc is full, a FATAL ERROR will be reported by LOTUS 123. If you were saving to floppy disc, insert a new formatted data disc and press ALT and Z to resume work. On a hard disc system, make room on your hard disc or save to floppy disc by using LOTUS 123 commands to change the current directory name to the floppy disc drive.

PRINT

The PRINT command will produce a printed copy of the data used in the Summary Results panels, along with the Summary Results themselves, the grain equivalents and the current BMR equations.

Make sure your printer is switched on and is on-line before issuing the PRINT command. If you erroneously give the PRINT command when no printer is attached, your system may hang, i.e. the computer no longer responds to commands. You will then have to restart your computer and start over again.

FILE

This option can be used to save to disc a file which contains your output summary table of energy requirements. In other words, you may save to file the same pages as the PRINT command prints on the printer. This file will have the extension .PRN and can be incorporated in a word processed document. You will be prompted
> **Name of results file:**

Give a meaningful name with up to eight letters so that you will easily remember the contents of the file.

GRAPH

If your video permits graphics and LOTUS 123 has been correctly configured for the display adapter you are using, this command will display graphs on your screen. The graphs use the data from the Work Area of the spreadsheet and therefore graphs should be requested only when the calculation is complete, WITH ALL NECESSARY DATA RETRIEVED. Graphs may be printed from *some* computers by pressing the PRINT SCREEN key, otherwise graphs may be saved to a file and printed later using the LOTUS 123 PRINTGRAPH command.

   Five graphs and SAVE and QUIT options are available from the following sub-menu:
POPULATION   WEIGHT   HEIGHT   INDIVIDUAL   TOTAL   SAVE   QUIT

GRAPH–POPULATION

This graph shows the age distribution of the population.

GRAPH–WEIGHT

The weight distribution of the population is shown.

GRAPH–HEIGHT

The height distribution of the population is given.

GRAPH–INDIVIDUAL

Energy needs in kilocalories are given for an average individual in each of the age and sex groups used in ENREQ. The Y axis is in thousands of kilocalories.

GRAPH–TOTAL

Total energy needs in kilocalories at country level by age and sex groups used are calculated by multiplying actual population numbers in each group by the average individual requirement. These are given in a graphic form with this command. Due to space limits, the Y axis is in thousands of millions of kilocalories.

GRAPH–SAVE

In order for a graph to be printed if the Print Screen function does not work on your PC, you must save the graph to a file, exit ENREQ and LOTUS 123 and use LOTUS PRINT–GRAPH. ENREQ's GRAPH–SAVE option asks

Name of graph to save

Supply a meaningful name. This graph file is then saved as a LOTUS 123 graph (.PIC) file. See the LOTUS 123 manual on how to print LOTUS graphs.

BREAK

To leave ENREQ but stay in the LOTUS 123 environment and move around the ENREQ spreadsheet with the arrow keys and the PAGE UP and PAGE DOWN keys to inspect the databases or the program itself, use this command. When you have finished, quit LOTUS 123 by typing

/QY

QUIT

To finish using ENREQ and to quit LOTUS 123, choose this option. You will be asked

Save current worksheet (y/n)?

Always say no unless (1) you want to save the entire spreadsheet as it is currently, with certain databases loaded and calculations on screen and (2) you have *not* already saved them using the SAVE or FILE command. Not saving the worksheet here does *not* mean that you lose the original ENREQ version because the program is already on disc.

If you press ⟨ESC⟩ instead of typing a y or n, it will have the same effect as typing n. Whether your answer is negative or positive, you will be returned to the MS–DOS environment.

ENREQ country groupings and spellings

Group 1
Afghanistan
Algeria
Angola
Bahrain
Benin
Botswana
Burkina Faso
Burundi
Cameroon
Cape verde
CAR
Chad
Comoros
Congo
Cote d'Ivoire
Egypt
Eq Guinea
Ethiopia
Gabon
Gambia
Ghana
Guinea
Guinea-Bissau
Iran IR
Iraq
Jordan
Kenya
Kuwait
Lebanon
Lesotho
Liberia
Libya
Madagascar
Malawi
Mali
Mauritania
Mauritius
Morocco
Mozambique
Namibia
Niger
Nigeria
Oman
Qatar
Rwanda
Saudi Arabia
Senegal
Sierra Leone
Somalia

Sudan
Swaziland
Syria
Tanzania
Togo
Tunisia
Uganda
United Arab Emirates
Yemen AR
Yemen PDR
Zaire
Zambia
Zimbabwe

Group 2
Australia
Bangladesh
Bhutan
Burma
China
Dem Kampuchea
Fiji
India
Indonesia
Japan
Korea DPR
Korean Republic
Laos
Malaysia
Mongolia
Nepal
New Zealand
Pakistan
Philippines
PNG
Sri Lanka
Thailand
Viet Nam

Group 3
Albania
Austria
Belgium
Bulgaria
Canada
Cyprus
Czechoslovakia
Denmark
Finland

France
Germany FR
Greece
Hungary
Iceland
Ireland
Israel
Italy
Luxembourg
Malta
Netherlands
Norway
Poland
Portugal
Romania
Spain
Sweden
Switzerland
Turkey
UK
USA
Yugoslavia

Group 4
Argentina
Barbados
Bolivia
Brazil
Chile
Colombia
Costa Rica
Cuba
Dominican Rep
Ecuador
El Salvador
Guatemala
Guyana
Haiti
Honduras
Jamaica
Mexico
Nicaragua
Panama
Paraguay
Peru
Suriname
Trinidad & Tobago
Uruguay
Venezuela

Users familiar with LOTUS 123 might wish to 'explore' the ENREQ spreadsheet and its program. To do this, ENREQ must be loaded into LOTUS 123, but not activated. If you are running ENREQ, quit by choosing QUIT from each successive menu until you reach the Main menu where you choose BREAK to go to the LOTUS 123 environment. Press the HOME key in order to go to the top left corner of the spreadsheet. This will bring you to the title page.

You may move from one screen to the next horizontally by pressing the Control key, and keeping it pressed, then press the right or left arrow keys. Moving from the title page to the right will bring the following screens in succession:

Program
area
$\begin{cases}\end{cases}$
Title page
Range names
Page1 Display area (printed together with PAGE2 when PRINT
    command issued)
Work area
Page2 Results panels

Lookup
Tables
$\begin{cases}\end{cases}$
NCHS wt/ht 0–9+ yrs
Baldwin wt/ht 10–17+ yrs
Fogarty wt/ht for adults

ENREQ
database
$\begin{cases}\end{cases}$
Data selection criteria
Population database (5 horizontal screens)
Weight database (4 horizontal screens)
Height database (4 horizontal screens)
BMR equations
Child allowance database (2 horizontal screens)
Physical activity levels database (2 horizontal screens)

The following diagram shows the various sections of ENREQ and their relative positions in the spreadsheet.

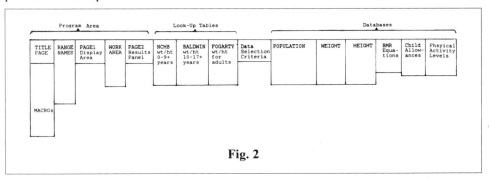

Fig. 2

Move vertically with the PAGE UP and PAGE DOWN. Moving PAGE DOWN from the title page will take you to the Macro definitions, which constitute the ENREQ program.

Another way of moving around the ENREQ spreadsheet is to use the LOTUS 123 key GO TO (F5) followed by the name of the section ⟨CR⟩ (in LOTUS 123 terminology, a RANGE NAME). Main sections and their contents are given below:

ALLOWANCE   Intake values for males and females 0–9+ years (requirement, desirable and infection, allowances, kcal/kg)

BALDWIN	Reference weights for actual heights (adolescents 10–17+ years)
BMR	Equation to predict BMR from weight
CRITERIA	Selection criteria for extracting records from the database
FOGARTY	Reference weights for actual heights of adults
HEIGHT	Height distance
MACROS	Macro definitions for spreadsheet. Of interest to those wishing to modify or examine the ENREQ program
PAGE1	Display of intermediate results of ENREQ calculation
PAGE2	Summary of the results displayed on PAGE1 together with grain equivalents and BMR equations. Both PAGE1 and PAGE2 are printed when the PRINT command is issued
PAL	Daily average physical activity levels for adolescents (10–17+ years) and adults
POPULATION	Population database including number of pregnant women
RANGES	Range names used in macros (Note: if new data are added to the databases, some of these become incorrect)
REFERENCE	Standard weights for actual heights (children 0–9+, and adolescents 10–17+ years)
TITLE	Title page
WEIGHT	Weight database

References

[1]FAO/WHO/UNU, 1985. Energy and Protein Requirements. Report of a Joint Expert Consultation. Technical Report Series No. 724, WHO, Geneva.

[2]Schofield, W. N., Schofield, C., and James, W. P. T., 1985. Basal Metabolic Rate: Review and Prediction. Human Nutrition: Clinical Nutrition, vol. 39C, suppl. 1,1–96.

Index

economic factors, effects on food
    availability 12–13
elasticity, nutritional 3
elderly, assessing energy allowances 98,
    100
emergency feeding 97
  energy requirements 97
  food needs 98
  see also food aid
employment statistics 55
energy
  constancy 35
  metabolism 37–8
  storage 35–7
energy allowances 5–6, 17–18, 67–82
  adults and adolescents 68, 73–80
  children 68–73
  effects of ageing 78–80
  factors affecting estimates 83–90
    actual versus desirable body weight
        83–6
    desirable weight and activity
        allowances 76–7
    interactions between 88–90
    population increases 86–8
    population structures 86–7
    urbanization 88–9
  impact on national energy
      needs 32–3
  lactating mothers 81–2
  in pregnancy 80–1
  special groups 98–100
  see also energy requirements
energy balance
  fluctuations 39–40
  principles 35–41
energy expenditure
  adjustment to semi-starvation 40–1,
      91–2
  assessment 41
  components 36–8, 43
  constancy 38–9
  of physical activities, simplifying
      estimates of 47–8, 50–7
  simplifying components of 42–4
energy imbalance, adaptation to
      40–1
energy intake 36
  effects of fluctuations 39
  low, *see* low energy intake
  metabolic adaptation to changes in
      37–8, 91, 94–5
  urbanization and 11
energy requirements 1–2, 17–18, 67
  calorie source and 5
  children 69
  climate and 3
  food supply patterns and 3–5
  of a group/population 2–3

1973 versus 1985 FAO
    recommendations 31–2
  basing estimates on energy
      expenditure 41
  in emergencies 97
  impact of urbanization 58–62
  principal determinants 32–4
  simplest approach to estimating
      26–30
  simplified calculation 23–5
  individual 101
  individual variations 96
  maintenance, *see* maintenance
      requirements
  nutritional elasticity and 3
  physical activity level (PAL) and
      95–6
  population structures and 62–5
  principles 35–41
  survival 95–6
  see also energy allowances; food
      needs
exercise, intensive 50–1

FAO reference weight for height
    equations 127
farm systems analysis 10–11
fat consumption 11
fat reserves, body 35–6
Fogarty reference weight for height
    values 128
food aid
  assessing need for 13–16
  regional deficits 15–16
  see also emergency feeding
food availability (supply)
  assessment 13–15, 21–2
  effects of structural adjustments
      12–13
  energy requirements and 3–5
food balance sheet 21–3
food costs, effects on food availability
    12–13
food demand 2
  elasticity 3
  impact of urbanization 11–12
food distribution systems 4
food intake, thermic response to 37, 40,
    43–4
food losses 5
  household 20–2
  in storage and distribution 20–2
food needs
  for emergency feeding 98
  nutrition-based method of assessment
      14–16, 23
  status quo method of assessment
      13–14, 16, 23